Microbes and Men

by the same author:
Tongues of Conscience
Marie Curie

ROBERT REID

MICROBES
AND MEN

British Broadcasting Corporation

Published by the British Broadcasting Corporation
35 Marylebone High Street, London W1M 4AA

ISBN 0 563 12469 5

First published 1974
© Robert Reid 1974

Set in Monophoto Imprint
and printed in England
by Jolly & Barber Ltd,
Rugby, Warwickshire.

OVERLEAF: Some members of the staff, St Mary's Hospital, shortly before the discovery of penicillin. *From left to right :* J. Freeman, L. Colebrook, A. E. Wright, A. Fleming, C. B. Dyson, J. Matthews, C. G. Schoneboom, F. Ridley

CONTENTS

FOREWORD

That percipient man J. D. Bernal, in what is far from being his best book, says that first-rate books on science are written by geniuses and are never understood at the time. Second-rate books on science (of which Bernal unconvincingly claimed to have read all) are, he said, written by ordinary people and understood by ordinary people. In that sense a book called *Microbe Hunters* by Paul de Kruif can claim to have been a first-rate, second-rate book on science since the number of ordinary people who read it, and were influenced by it in the ten or twenty years after its publication, was enormous. Dr Alfred Alexander first drew my attention to de Kruif's work and for this I am very grateful. It gave me the kernel of the idea from which developed the television series which this present book accompanies. And though otherwise this short work in no way derives from de Kruif, I think it only appropriate to pay tribute here to an honourable scientific populariser, who caused a generation of young people to look in wonder at a scientific subject which without his influence would have passed them by unnoticed and unsung.

Besides de Kruif and Dr Alexander, I am just as grateful to Professor D. C. Burke, Peter Campbell, Dr W. D. Foster, Peter Grose, John Knowler, Dr R. Lancaster, Dr S. Selwyn and Professor R. Whittenbury for their comments on parts of the manuscript. I am also thankful for material provided by the television series production team, Peter Goodchild and his assistants Fiona Holmes and Graham Massey. Few who have not worked in television can appreciate the prodigious trouble and the immense number of hours spent in trying to achieve accuracy in television documentaries and, in the case of dramatic reconstruction, the best approximation to truth. This introduction gives me the opportunity to point out and praise the fact.

Burying victims of the Plague of London of 1665 in multiple graves

CHAPTER ONE

Lady Mary Wortley Montagu swept into Turkey in 1717. Her husband, Edward, was the new Ambassador in Constantinople. At 23 she had eloped with him and now, a more discerning 28, she was beginning to discover her mistake with the tedious and unaccountably miserly career diplomat.

Turkey too could have been tedious, but Lady Mary had no intention of letting it be so. Two hundred years ahead of her time, she was an active feminist, had secured her own liberation by virtue of her innate intelligence and self-education, and was inclined to take on men on their own terms as well as derive whatever advantages and pleasures she could from being a woman. She had a tongue like a viper and a pen like a razor and could make her presence felt in any society she found herself. And she had the sexual attraction and experience to have her literary excesses and her cutting wit applauded when others, physically less well-endowed, might not have been tolerated. For example, she carefully observed the new German King of England at the court of St James and, just as closely, watched his mistress, whom she described as being 'duller than himself and consequently did not find out that he was so'. The observation was repeated with glee both in and out of court.

Lady Mary herself intended to experience life to the full, and left an account of that life in diaries and letters which she had no doubt posterity would read. Some were so intimately revealing that posterity, as represented by a few prudish individuals, destroyed them. Alexander Pope, having observed Lady Mary's generous proportions and behaviour, described her as an Eve who tasted not one apple, but robbed the whole tree.

In Turkey Lady Mary took every precaution not to be judged the typical English resident who spent most of the time behind the Embassy shutters, then returned to Europe with grossly exaggerated travellers' tales from the barbarous Near East. She found the Turks interesting and their women civilised. So much so that she posed for her portrait in Turkish dress (but nevertheless retained the fashions of the English court and made sure that posterity would also take note of her well-shaped breasts, amply revealed). She spent no time decrying the habits of the natives. She was quick to observe and record political and social customs of which she approved. She soon noticed the justice of married women's property rights in Turkey; in particular that on divorce women kept their own money and had maintenance from the husband.

She was highly interested in matters of health and medicine. For example she recorded how Turkish women, after childbirth, saw visitors

9

on the same day as their delivery, and were up and about in their jewels and finery returning calls within a fortnight. It was 200 years before this sort of post-natal behaviour and a similar interpretation of the Married Women's Property Act came into being in Britain.

The delights of the Turkish bath were a revelation and through experience she formed the opinion that 'if it was the fashion to go naked, the face would be hardly observed'. As it happened Lady Mary's face and beauty, together with her pen, were her most important assets in Whig society. Like everybody else in that society she had reason to fear that her beauty might be short-lived. Smallpox is a word which now occasionally raises a flutter of public interest simply because, in Western countries, the disease is a rarity. That it should have become so is almost unbelievable when we look at its effects on eighteenth-century Europe. In those hundred years upwards of fifty million people died in Europe alone from this one disease. In certain towns in England seventy-five per cent of the population were said to be affected by it, and one in four to have died from it. Its physical effects were awful. Along with a fever, blisters of pus and running sores would form, sometimes together with red rashes on the body and limbs. In bad cases the pustules would cover the whole of the face and eventually form scabs which would turn to deep ugly scars. The distress caused by the disease was appalling. If a child at its mother's breast contracted smallpox, and if the mother herself had not yet suffered from it, she had the choice of risking her own life by continuing to feed the baby, or risking her baby's life by trying to wean it too early.

At some time in her life Lady Mary Wortley Montagu had suffered mildly from smallpox and it had dimmed her beauty, although not so badly that she could not hide its worst effects. Having once suffered, she knew she could observe the disease from close quarters with little fear, and in Turkey she took advantage of her immunity and with her customary attention to detail made a full record of what she saw in Constantinople.

Every September, when the heat of the Turkish summer had abated, she watched a group of professional old women do the rounds of houses in Constantinople. Their purpose was to practise a well-tried piece of preventive medicine. Families and friends would gather together at the houses and undergo the ritual of *ingrafting* at the hands of the old women. Each crone would carry in a nutshell a small sample of pus she had collected from a victim of a mild attack of smallpox. With a large needle she would quickly and fairly painlessly scratch open a vein on the limb of her customer. She would then dip her needle into the pus in the nutshell, smear it on to the open vein, cover the wound with a piece of nutshell and bind it.

Lady Mary noted the effects on children who had been treated by the old women. They would

play together all the rest of the day, and are in perfect health to the eighth. Then the fever begins to seize them, and they keep their beds two days, very seldom three . . . and in eight days' time they are as well as before their illness . . . Every year thousands undergo this operation; and the French embassador says pleasantly, that they take the smallpox here by way of diversion, as they take the waters in other countries. There is no example of any one that has died in it; and you may believe I am very well satisfied of the safety of this experiment, since I intend to try it on my dear little son.

True to her word, that year, 1718, Lady Mary supervised her first medical experiment on her own child. Edward Wortley Montagu was the first recorded Englishman to experience this form of inoculation against smallpox. He survived, and not only was he protected against the dreaded disease, but the scars left by the operation presented an added bonus to his mother. Edward used to make a habit of running away from his school, Westminster. On one occasion a reward of £20 was offered to whomever found him. The advertisement to retrieve Edward gave unmistakable means of identification. There were 'two marks by which he is easily known; viz., in the back of each arm, about two or three inches above the wrist, a small roundish scar, less than a silver penny, like a large mark of the smallpox'.

Smallpox was so common in Europe at the time that Mary considered it nothing short of her patriotic duty to make 'this useful invention' fashionable in England. Back home in 1721 Dr Charles Maitland, who had been physician to the English mission during her stay in Constantinople, began to practise inoculation under her patronage. Immediately Lady Mary's 'heathen' ways came under fire from both the established medical profession and the church. But she had both the persistence and, more important, the contacts to further her ends: the widespread inoculation of the population of England against smallpox.

Her most imposing contact was at the Court of King George I. Notwithstanding the fact that, like Pope before him, the Prince of Wales had eyed her body appreciatively, she formed a friendship with Caroline, Princess of Wales. She had also begun to reformulate her opinions of that 'honest blockhead', the King. Through the good offices of the Court, and with the help of the very distinguished secretary of the Royal Society, Sir Hans Sloane, she arranged as improper a series

Lady Mary Wortley Montagu

of medical experiments as has ever been conceived by man, or in this case, by woman. Seven condemned criminals from Newgate prison were given the option of the gallows or of submitting to Lady Mary's newly imported smallpox inoculation. The latter choice might have led to one of several alternative results; they could have their freedom and with it either immunity from smallpox, disfigurement, or death from the disease. In this macabre form of roulette – which depended ultimately on Lady Mary's accuracy in reproducing what she had seen in Turkey – the criminals chose inoculation. They survived with nothing worse than their inoculation scars, and their freedom.

The experiment was repeated one step up the scale of undesirable human life on half a dozen orphan children, who survived. By 1722 the King was sufficiently impressed by Lady Mary's successes to be persuaded to have two of his grandchildren inoculated. The ingrafted scar was now a royal imprimature and the fashionable success of Lady Mary's oriental discovery was assured. She became a celebrity in her own time and secured for herself a niche in medical history.

She also made exaggerated claims for her method. She declared sweepingly, 'I know nobody that has hitherto repented the operation'. In fact there were many deaths through smallpox inoculation. And even though the number of deaths was far outnumbered by the lives saved, the method was soon tainted by fear and it was neither widely, nor without controversy, adopted during the years that followed.

Lady Mary Wortley Montagu was no scientist (quite apart from the fact that the word had not yet been invented), and would not have laid claim to be so. Her observation of smallpox inoculation was far from original; a similar form of inoculation was probably in existence in more barbaric parts of Britain – in Scotland and Wales – when she introduced it as a novelty to England. But she had applied certain scientific principles to her observation. She had, like others before her, thought up a theory to link inoculation of mild smallpox with immunity from smallpox, and had devised experiments, immoral as they undoubtedly were, to test her theory. Finally she had published her results: broadcast with a fanfare would perhaps be a better description.

Her flair for personal publicity was an important ingredient in her successful impact on eighteenth-century scientific thinking. Without it, without royal patronage and a fashionable following, the discovery might have remained as hidden as it was before Lady Mary appeared on the Turkish scene. In the tradition of the English amateur natural philosopher, the tradition of Bacon and of Boyle, she had made her contribution to science well. Today that element of publicity is as important as it was in 1722. Appropriate publication is still vital to the process of thrusting a discovery or piece of creativity into the current of scientific thought.

In the 250 years since the children of the Royal House of Hanover were put to the risk of inoculation, the vast strides that have been taken to conquer the once unopposable and terrible smallpox have been phenomenal. In 1973 just one isolated case of the disease contracted in a London medical laboratory caused a panic, a scandal and a public inquiry. In the late seventeenth century one doctor, even when the plague was not particularly virulent, blandly estimated that the annual mortality in London due to smallpox was around 3000. His was probably an underestimate. Between these two dates and these two contrasting attitudes lie much suffering, many scientific and medical experiments, and many individual contributions to the corporate scientific verification of a theory of disease.

This theory, the germ theory, simple in its concept, was built up like a mosaic from fragments of scientific discovery. How this was done has a great deal to tell about the process of science, about how scientists operate individually and corporately, and about the nature of scientific creativity. When the germ theory began to emerge from the mists of antique medical practice there was little to choose between doctors and old wives and their tales. They had a mystique of Hippocratic tradition behind which they could hide when they were in ignorance. A patient knew that the man he was employing had sworn the Hippocratic Oath. Other than that there was no guarantee that the results of the consultancy would be any better than the work of an apothecary, a priest or a butcher. The germ theory turned medicine from a frequently hypocritical and a frequently pseudo-rational activity into a science-based practice. The outcome of its verification was socially and politically as powerful and far-reaching as the applied results of any other experiment in physics, chemistry or biology, before or since.

CHAPTER TWO

Scientifically, Lady Mary Wortley Montagu's dilettante discovery was less interesting than several other contemporary experiments bearing on the understanding of the nature of disease. Her contribution to scientific knowledge was important only because she made her observations part of the common intellectual currency of the eighteenth century. However, her work also shows that, by the beginning of that century, the nature of contagion was, in a very limited sense, understood.

There was a clouded comprehension, for example, of how smallpox could be transferred from one human being to another, carried on the end of a needle. This kind of elementary knowledge relating to other diseases had been available for centuries, as Biblical records and early taboos show. Religious laws based on such knowledge condemned the leper to a lonely exile of progressive suffering. The very word *quarantine* derives from the Italian 'quarantenaria', and originated in the fourteenth century when the Black Death was ravaging Europe and sea voyagers into Ragusa harbour in Sicily had to wait forty days before being allowed to enter the city.

In England in the sixteenth century taint on the monarch's purity was explained away by the application of some rudimentary understanding of contagion. The origin of Henry VIII's venereal disease was attributed to Wolsey, and the fact that the ill-fated Cardinal had been whispering in his King's ear. Thus were the purveyors of secrets punished, and thus was the history of a powerful nation disastrously influenced by its leader's impaired judgement.

Syphilis, that gentle, sibilant name which, in its Greek origin, means hideousness, was what the Greek shepherd hero in a poem by Girolamo Fracastoro was called. In his narrative the youth, Syphilis, was inflicted with the terrible body-rotting disease by the insulted god of medicine, Apollo. Fracastoro followed his poem, which had a wide popularity in European literary circles, with a factual work, *De Contagione*. This scientific treatise contained the elements of a theory of disease. In it Fracastoro guessed that particles of some sort ('seminaria'), which were self-reproductive and had a lifetime of perhaps two or three years, were responsible for contagious illnesses. Contagion could be by direct contact, through materials touched by infected individuals, or by passage over a distance through the air.

Such was Fracastoro's theory: but he made no attempt to verify it by experiment. So by the pragmatic Lady Mary Wortley Montagu's time (Lady Mary having rushed into experiment without a glance behind her), there was still no better understanding than there had been in

Biblical times of how – by what mechanism – contagious diseases were transmitted. It was known with some uncertainty, and with fatal consequences for the victims of that uncertainty, that it was possible to confer immunity from smallpox. The method was well known in some oriental countries and, by the end of the eighteenth century, in England also. It was certainly well known, through personal experience, to a West Country doctor, Edward Jenner.

Born in 1749, Jenner was not only familiar with the technique, he had suffered some of the eighteenth-century European refinements. In childhood, during a bad epidemic in his native county of Gloucestershire, he had been inoculated with smallpox and had suffered painfully at the hands of doctors who had bound Lady Mary Wortley Montagu's method in wreaths of unnecessary and dangerous medical mystique. The young boy had been bled to make sure 'his blood was fine', purged until he was emaciated, and kept on an inadequate diet. In spite of his doctors, he recovered, fully protected from smallpox, and sufficiently impressed by the method to use it himself when he took up his country practice. By the time he was adult, although it had been stripped of some of its charlatan trimmings, it still had to be applied with rustic vigour and unemotional concern for the results. Jenner saw many cases of smallpox inoculation turn into severe cases of smallpox itself.

But Jenner was not an unemotional man. From childhood he had hypomanic moods, varying from periods of high mental activity and vivacity to dark moods of melancholy and introspection. It would be difficult to find a statistician who would base his reputation on a theory, of creativity related to type, but nevertheless Jenner's was a character which is frequently found in the lists of people with creative scientific temperaments. Certainly, there are several others, with similar temperaments to Jenner, to be found associated with the germ theory.

Jenner belonged to that middle-English country class, solidly sited in the Vale of Berkeley, watched over by a God well established in his heaven, and stabilised by many families, like Jenner's with large proportions of their members either parsons, or married to parsons.

He had every wish to make his mark in the accepted scientific tradition of the class that bred him: scholarly, but amateur. His most profound interest was bird-watching and he went to some trouble to register the fact that it was the robin and not the lark which heralded day-break, and to determine the order of appearance of the songs of the raven, the jay, the swift and others. But at the centre of Edward's passion was the cuckoo. He spent many happy days, escaping his melancholia, in observing this bird's victims and trying to catch the hedge-sparrow in the unnatural act of throwing their young from their nest so that the intruder could survive. Instead he discovered that the real murderer was the young cuckoo itself, who, using a hollow in its back as a spoon, pitched out the rightful dwellers. His paper on the natural history of the cuckoo, addressed to the famous anatomist, John Hunter, earned Jenner election to a Fellowship of the Royal Society.

In 1770 Hunter had taken Jenner as his first house-pupil at St George's Hospital at Hyde Park Corner (then still within sight of the open countryside). Hunter is deservedly remembered for his successful attempts to turn his profession away from techniques little better than butchery, by introducing simple biological theory into surgery. But there was always a clear practical purpose as the aim of Hunter's revolutionary theoretical approach to medicine. The title of his last work, *A Treatise on the Blood, Inflammation and Gun-shot Wounds*, briefly but adequately illustrates his practical professional approach. His first pupil inherited the clear principles of his method.

When Jenner returned to the Vale of Berkeley to try out the skills he had acquired from Hunter, he did so in the casual and relaxed fashion which was the tradition of the country gentleman doctor. In 1788 there was an epidemic of smallpox in Gloucestershire. Fear of the disease brought back the popularity of inoculation, as it always did during an epidemic, in spite of the often tragic shortcomings of the method. Jenner administered inoculations on his daily rounds. However, he noticed some peculiar reactions to his efforts in parts of the county where cows were herded. Patients who had had what was locally called cowpox – a disease which affected cows' teats and subsequently the hands of their milkers – remained uninfected in spite of all Jenner's efforts to give them mild attacks of smallpox.

Jenner was well aware of the old wives' tale common among countryfolk, that those who had suffered from cowpox were resistant to smallpox. He had first heard the tale while working his apprenticeship in a country practice in Sodbury. A dairymaid he had come across, with a skin infection of pustules, had spoken up with the confidence of experience and announced that what she was seeking a cure for could not be smallpox, since she had already had cowpox. The remark stayed in Jenner's mind. He had no reason to act on it quickly. Why should he? This, and countless other old wives' tales like it, had been around Gloucestershire for centuries. At the leisurely pace that he had marked his observation of the cuckoo, he began to observe outbreaks of cowpox. He followed case histories of the disease just as he had followed the case history of the parasitised hedge-sparrow. Cowpox appeared on the teats

17

of cows as inflamed pustules and quickly spread to the whole herd. Soon afterwards milkmaids and milkmen would develop sores on their fingers' ends and finger joints. The sores would spread to other parts of the body and a fever would set in which would die down after a few days. At first Jenner had a theory that this was a form of smallpox, that it started as an inflammation in the heels of horses called *grease*, and that it spread to cattle, then to humans as smallpox.

Over the years he observed dozens of cases. His first recorded case was that of Joseph Merret, under-gardener to the Earl of Berkeley. In 1770 Merret had had cowpox. When Jenner tried to inoculate him against smallpox in 1795, even after several attempts, it was possible to raise only a small swelling on his arm. When Merret's family was badly hit by smallpox during one epidemic, Jenner noticed that Merret remained quite unaffected, although he continued to live among and tend his family.

Jenner was also consulted by 'a respectable Gentlewoman', and observed that cowpox was no respecter of classes, unlike Jenner himself, who discreetly referred to his patient as 'Mrs H____'. Mrs H had contracted her cowpox by handling milking vessels used by her farm servants. Not only her hand, but her nose 'became inflamed and very much swoln'. Then, when smallpox swept the countryside and Mrs H was in the thick of a family infection and tending one relative who died from the disease, Jenner observed she survived unscathed. When he tried to inoculate her she showed little reaction.

Jenner, though still not a man in a hurry, was now ready to act. It was in 1796, a quarter of a century after he first heard the dairy maid's tale, that he set out to experiment. Unlike Lady Mary Wortley Montagu in her experimental activities, as a doctor, he was bound by a code of conduct: the Hippocratic Oath. This solemn set of rules is, as it happens, totally ambiguous in interpretation, particularly where it refers to ethical behaviour towards a patient. If he thought of his oath at all, Jenner liberally interpreted its clause, 'I will go to help the sick and never with the intention of doing harm or injury'. Some might say he more than adequately overstepped its bounds.

The human subject he chose on 14 May 1796 was a villager well below the class of Mrs H. In Jenner's words: 'I selected a healthy boy, about eight years old, for the purpose of inoculation for the Cow Pox. The matter was taken from a sore on the hand of a dairymaid who was infected by her master's cows'. Jenner took the arm of the boy, James Phipps, on it he made two small cuts about half an inch long, and smeared pus in them. After a week Phipps developed a slight fever which quickly cleared. He was left with nothing worse than a pair of scabs on his arm.

On 1 July Jenner then took some smallpox 'matter' and proceeded, as he told a friend, 'to the most delightful part of my story'. He inoculated the boy several times on both arms. Triumphantly for Jenner, and fortunately for James Phipps, the smallpox had no effect. Chance selection took Phipps into history. He was subsequently inoculated with smallpox more than twenty times, without any untoward effect. Jenner had invented vaccination.

Jenner continued his experiments in a random fashion depending on the local outbreaks of cowpox and the availability of pustules on the hands of farmhands for his vaccine. The elation of success pushed him into a flurry of activity which life had never demanded of him before. But his experimental successes were by no means an unbroken line. On 16 March 1798 he inoculated a five-year-old child from the local parish poor-house, John Baker. Jenner used matter from the hand of a sick farm worker who, Jenner believed, had contracted his infection from the heels of the horses he tended. John Baker died, according to Jenner, 'having felt the effects of a contagious fever in a workhouse soon after this experiment was made'.

Jenner's work was muddled and risky, but after twenty-five years of inconsequential observations of natural history as his outlet for his depressions, he now knew that he had carried out experiments of real substance. Jenner's familiarity with the activities of natural historians had also made him familiar with the process of publication of scientific observations. At for him an unprecedented speed, he sent a paper to the Royal Society reporting his cowpox investigations. The first reaction of the Fellows of that Society, conscious of Jenner's natural historical success with the cuckoo, were to advise that, 'he ought not to risk his reputation by presenting . . . anything which appeared so much at variance with established knowledge, and withal so incredible'. But it was precisely because this work ran counter to established knowledge that it was important. Eventually, however, a committee of Jenner's friends urged him to publish privately. In June 1798, *An Inquiry into the Causes and Effect of the Variolae Vaccinae* appeared. Its dedication was from Lucretius: 'Quid nobis certius ipsis sensibus esse potest, quo vera ac falsa notemus' (what can we have that is more reliable than our senses to distinguish between truth and falsehood?). Jenner's work had entered the main-stream of scientific communication. Widespread awareness of it was assured, even though he had had to pay for the thing himself. On sale at seven shillings and sixpence a copy, it was good value to mankind.

Within a few years vaccination was commonplace. Jenner had found how to preserve active vaccine and his method had spread across Europe. The first Russian child to be given immunity from smallpox was christened Vaccinoff and given a pension for life; James Phipps had to make do with a free cottage.

But vaccination was neither without its detractors nor its tragedies. Some practitioners attempting the method succeeded in giving their patients serious cowpox infection. 'Perhaps,' said Jenner (who was no clearer than anyone else about the mechanism which made his method a success), 'the difference we perceive may be owing to some variety in the mode of action of the virus upon the skin of those who breathe the air of London and those who live in the country.' And Jenner had to put up with a certain amount of mockery. Wasn't it possible, one wit asked, that humans inoculated with diseases of beasts would develop bestial characteristics? It was as scientific a riposte as some of Jenner's statements. In fact he offered such a variety of excuses for failures of the method that the Prince of Wales' own surgeon remarked that 'sometimes the cow was to blame, and sometimes the doctor'.

But Jenner got both fame and reward. In the House of Commons, which was being called upon to vote a reward to Jenner for making his discovery generally available, Admiral Berkeley took up his neighbourhood doctor's case in the debate and said:

Suppose it was proposed in this house to reward any man who saved the life of a fellow-creature with ten shillings I should be laughed at for the smallness of the sum, but small as it is, I should be contented with it, for if the statement of 40,000 deaths is true, and if this discovery prevents it, Dr Jenner would be entitled to £20,000 per annum.

In spite of the fact that there was a strong feeling that Jenner had already financially rewarded himself quite well enough from the increase in the vaccination practice he had started in Gloucestershire, he was given £10,000. Though, with a grudging hold on the nation's purse-strings, the Chancellor of the Exchequer felt that Jenner had already received the greatest reward any individual could receive: 'the unanimous approbation of the House of Commons'.

The motion to give Jenner his Treasury grant was only narrowly carried, by 59 votes to 56. One of the arguments that weighed heavily against him in the minds of the purse-conscious members of the House of Commons committee of inquiry was that his discovery of vaccination was not original; at the sound of jingling cash, others had appeared rapidly on the scene with claims to pre-date Jenner.

LEFT: Edward Jenner

BELOW RIGHT: Dairy maid's arm infected with cowpox. From Jenner's *Inquiry into the Causes and Effects of the Variolae Vaccinae*

BOTTOM: St George's Hospital, London, in 1745

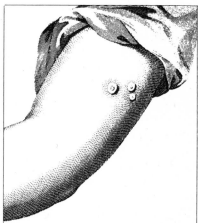

The COW POCK — or — the Wonderful Effects of the New Inoculation! — Vide. the Publications of ỹ Anti-Vaccine Society

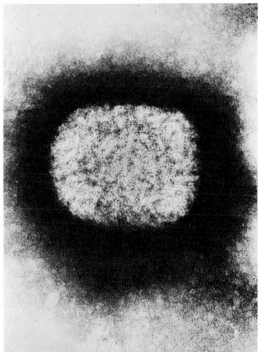

ABOVE: Smallpox inoculation as seen by Gillray in 1808

LEFT: Electron micrograph of smallpox virus magnified approximately 600,000 times

What the House of Commons committee was attempting was to pass judgement on scientific creativity. Was it true that Jenner was not the first to assemble the theory that cowpox can give immunity to smallpox? The answer is, yes. Unquestionably, in the terms that some of the members of the committee argued it, the benefactor of mankind could just as well have been the anonymous milkmaid of Sodbury who fed the idea to Jenner, and which, by chance, he preserved in his mind.

Chance is a word which is heard frequently in all accounts of scientific creativity. The impressive-sounding sentence, 'chance favours the prepared mind', or variants of it, has been attributed to any number of scientists, and to one particularly concerned with the germ theory of disease, Louis Pasteur. It recurs with meaningless regularity. Chance is the continuous, but random, occurrence of events in nature. Without a mind to interpret these events they therefore have no meaning. The Sodbury milkmaid had become aware of a series of random occurrences, made one observation in particular on herself, and had formed a theory. Jenner did no more until, crucially, he tested his theory. He did so on James Phipps in a way which today we would find ethically reprehensible; but without having done this, or some equivalent action, he would have been no more distinguished than the milkmaid.

The second factor which raised Jenner's work above the commonplace of folklore was that it was published in a manner that gave it meaning. He lived in a time when the methods of science were coming out of their Dark Ages. It was the time when the schemata and conclusions of Galileo, Newton and Lavoisier turned different aspects of discovery into fundamentally important words or phrases in the vast language, or hieroglyph, of science. As one word or phrase follows another, understanding grows. There is an added complication, in that greater understanding reveals vast new areas of the hieroglyph awaiting translation.

By her flair for personal publicity, Lady Mary Wortley Montagu had made her small contribution part of this progressive understanding, adding a word which continued an intriguing sentence. Jenner, because he was an amateur scientist familiar with the new and growing procedures whereby a scientific idea was announced and established, was able to make his contribution despite his comparative obscurity. The manner in which he published his work was fundamentally the same as that used by modern scientists.

There is no doubt that there was an establishment to uphold orthodox scientific knowledge. The Royal Societies of several European countries existed to do just that. What is characteristic about a piece of scientific creativity is that it changes the sense of the paragraphs that spell out the orthodoxy. It does not just carry the story forward; it sends them forcefully and irrevocably in another direction. Often the most revolutionary, influential and beautiful pieces of creativity are those which are performed contrary to establishmental or traditional methods.

The problem then is to sort the cranky from the valuable theory. The only way is to put the theory to the test: to devise experiments by which the theory might be falsified. If it is not falsified it will contribute to the vocabulary of science. If it is falsified, it is generally forgotten and left behind along with the contributions which were never published: never made available to the criticisms of orthodox science.

Jenner's work satisfied two of the criteria which qualified it as a piece of creativity to be admired; they are criteria which on the surface seem banal, and which cannot necessarily be applied to art: he tested his theory and he published his results. If Jenner had not devised a test, and if he had not published, his work would be sterile.

What motivated Jenner to make his creative contribution? Jenner would answer that he was moved by the possibility of doing good for mankind. Indeed, he wrote:

While the vaccine discovery was progressive, the joy I felt at the prospect before me of being the instrument destined to take away from the world one of its greatest calamities, blended with the fond hope of enjoying independence and domestic peace and happiness, was often so excessive that, in pursuing my favourite subject among the meadows I have sometimes found myself in a kind of reverie.

But this was written with success well in hand and with the doubtful benefit of many years of hindsight. Jenner spent a quarter of a century gestating a theory which, when he observed it to be very likely true, gave him enormous satisfaction as once the solution to the case of the murderous habits of the young cuckoo had done. Only after the event did the good it could do to mankind feed back on him and push him into establishing a thriving practice in inoculation. In the behaviourist sense of reward followed by a stimulus, this substantial reward was important to Jenner, and stimulated him to furthering his inquiries; of that there can be no doubt. But whether it contributed to the actual process of discovery is not, nor can be, proven. Any thoughts on mankind's good came *after* the creative act. Jenner's initial motivation, on available evidence, was that act of self-stimulation by curiosity leading to creativity, which differentiates man so powerfully from other creatures.

Sometimes, as he looked back on his discovery, Jenner, the countryman from the clerical family, felt the need to acknowledge the part played in it by God. But in his more morbid moods he put the matter differently. His way of acknowledging the part played by chance in the process of discovery was to describe himself as being one of the 'puppets danced about by wires that reach the skies'. And remind himself of the helplessness of his puppet role, he, the doctor who had secured immunity from one terrible disease, had to sit and watch his wife spit blood, she having contracted tuberculosis.

Jenner's contribution to the understanding of the language of contagious diseases was a small word in the face of what still needed to be translated. In London the death rate from tuberculosis was about one in six, and the rate was rising. Jenner had shown how a human could be made immune to just one ravaging disease, smallpox. But an understanding of the mechanism by which the disease spread itself, and therefore the beginnings of an understanding of how immunity to it operated, were nowhere in sight.

Ignaz Semmelweis. From a portrait by Abranyi Lajos in 1858

CHAPTER THREE

Until well into the twentieth century the death of a woman in or after childbirth was accepted as the recurring price of sexual intercourse. Childbed fever – puerperal fever – was a terrible blight on every lying-in hospital in Europe, particularly during the first half of the nineteenth century. The pattern the disease ran was cruel. On the third day after delivery the woman's pulse would rise, she would become prostrate with fever, develop a discharge from her womb, become delirious, have diarrhoea and shortly, as nothing could usually be done to stem the vicious progress of the disease, die.

The extent to which childbirth fever swept up its victims was vastly different in different hospitals. In 1856 a newly graduated young doctor, working his first year in the Paris *Maternité,* was shocked to discover that the overall deathrate for that year was 6 per cent. The fever struck in sudden, uncontrollable waves. During one of these periods he wrote, 'From the first of May to the 10th there were 32 cases of labour and we registered 31 deaths'. Belatedly, the hospital was closed.

If there were any useful lessons to be got from history, the *Maternité* had conspicuously failed to learn them. Seventy years earlier, in Manchester, Charles White had claimed to have radically reduced his cases of the fever by insisting on cleanliness, isolation of cases whenever they occurred, and by getting women off their beds within twenty-four hours of delivery. And ten years after that Alexander Gordon in Aberdeen had shown just how fatally easy it was for midwives to spread the infection to their patients.

But in most major cities of Europe the disease recurred with depressing persistence. Vienna of the 1840s had a particularly bad reputation in spite of the fact that it was the city which housed one of the most enlightened Medical Schools of Europe: the Vienna General Hospital. When Ignaz Semmelweis joined the lying-in clinic of this hospital as an obstetrical assistant in 1844 he knew its reputation for puerperal fever rate very well; that year almost one in ten pregnant women who entered it died.

Semmelweis, who came from an affluent Hungarian family to study in Vienna, had chosen medicine as a vocation, originally having travelled to Austria to read law. He was an emotional young man, and the condition of the women suffering from childbed fever moved him. The state of the wards which he had to walk, however, he probably accepted as commonplace; the primitive hospital management and facilities did not differ vastly from city to city. In the Vienna General, wards were crowded. The stench from the open sores of infection was appalling, and particularly so, the smell from the discharge from puerperal fever. But there were other sources of odours. It was customary for the closets, with their open sewers or buckets, to lead directly from the ward. In many cases – and the delivery room of the lying-in division in which Semmelweis worked was a case in point – autopsy rooms were joined by doors directly to the wards. Frequently students learned their medicine on the dead bodies of patients in the room next to which the patients had once lain. There was no universal tradition of cleanliness, either with respect to patients, practitioners or surroundings. A surgeon would not necessarily clean or change his apron between operations, and some would wear the bloodstains on it like medals, carrying them round from day to day, and from patient to patient. It was fashionable to tuck whip-cord, used for tying up arteries, in the buttonhole of the lapel of the coat. Wards were washed monthly or yearly according to nursing whim.

Semmelweis was a well-built, prematurely bald young man, already running to fat, who looked older than he was and whose behaviour had the authority of age. His authority was mixed with a certain dynamism which would qualify him for the twentieth-century definition of a Hungarian, as a man who goes into revolving doors behind you and comes out in front. When, quite soon after his arrival at the Vienna General, he began to question some of the techniques he saw in hospital practice, he was listened to with attention.

He was also a truculent young man with no evidence of tact. As a new-comer to the hospital, he had no hesitation in letting anybody, including his superiors on the teaching staff, know his conclusions on all matters relating to obstetrics. His methods of communication seriously handi-capped him throughout his life, particularly here in this hospital: the pride of the Austro-Hungarian Empire, and founded by the Emperor himself. Semmelweis was over-conscious of his own Hungarian origins, spoke German with a thick Hungarian accent, and never mastered the art of written German.

Semmelweis's immediate superior at the Vienna General, Klein, was professor of midwifery and director of the teaching clinic of the lying-in division. There were two clinics in the division. These had been formed when the hospital had been reorganised in 1840. The second, staffed by midwives, was run in the same way as the first clinic, except that medical students were not taught there. The existence of the two clinics was to make Semmelweis, first by implication then *de facto,* a virulent critic of Klein.

Semmelweis' first untoward observation was not merely unoriginal, it was shared by the vast majority of those who, as patients, had experience of one of the lying-in clinics; it was that the death-rate from childbed fever in the first clinic was far higher than that in the second. It was an opinion not generally shared by the rest of the medical teaching staff. Semmelweis had become convinced that the gossip among patients was true when he saw the depths of their fears of the first clinic. His easily stirred emotions were disturbed by several experiences. He watched cabbies driving their horses round and round the hospital square late at night, waiting until midnight struck, so that the heavily pregnant women inside need not be taken in to the hospital until the early minutes of the morning. The midwives admitted women in labour only on certain days of the week. Some women would prefer to give birth in the streets rather than be taken in to the first clinic. Working in that clinic Semmelweis himself had seen women with dry tongues, with high pulse rates and high temperatures, feigning good health in order to get an early discharge. And he had seen them get down on their knees to beg to be released from what he soon began to agree appeared to be a sentence to a higher chance of death.

There was even a young wives' tale as to what caused the death-rate in the teaching division; it was due, women said, to the examination by doctors. As he walked the wards, Semmelweis watched how obstetricians and midwives worked. Fundamentally he could see no differences, either in the way staff behaved, or in the organisation of the two clinics. But there was a way in which he could investigate. The chance organisation of the lying-in division itself presented him with an experimental technique which could be applied to medicine as it had never been applied before. The arrangement of the divisions into two parts provided Semmelweis with a control group of patients in one clinic against which he could measure the effects of any changes of conditions in the other. Even better, the Vienna General, being a Teutonically efficiently documented establishment, kept an exact record of admissions, deaths and causes of deaths, to which Semmelweis had access. He collected the statistics of childbed mortality for the two years to 1846 that he had worked in the hospital, and for the four previous years. The results were unambiguous. In the teaching clinic in those six years 1989 Viennese women had died: ten per cent of the total admissions. In the midwives' clinic the figure was nearer three per cent. However bad the figures for the midwives' clinic had been, in no single year had they shown any sign of being as devastating as those in the teaching clinic.

This discovery of a control group, and the application of simple statistical methods, was a landmark in the understanding of the nature of disease. So far Semmelweis, the large blunt Hungarian among the Austrians, now so frequently depressed by the deaths around him, had only used the technique with already existing data to emphasise to those with whom, and for whom, he was working, that there was something seriously wrong in their system. But the technique was such that he could easily use it to try to discover how puerperal fever was so effectively killing-off so many young women.

There were almost as many theories of the origin of the disease as there were deaths. It was said to occur frequently in cases of still-births, to depend on whether a woman was married or single, to be influenced by the weather or to be caused by a mysterious halo clinging to the obstetrician. For many years obstetricians had taught that it was a fever of the milk, and one French worker even claimed to have found milk in the peritoneal cavity. Some practitioners had decided that diseases such as scarlet fever, measles, smallpox and erysipelas were related to childbed fever and that one could turn into the other.

How the disease spread was another tale of ignorance. Some medical schools had it that the disease was contagious, but there was no useful theory as to how the contagion spread. Much mumbo-jumbo was talked about 'atmospheric telluric influence'. Looking back over more than a hundred years it is easy to despise this terrible replacement of ignorance by ignorant phrases. Nevertheless, this sort of phrase is not far removed from some in daily use in modern medicine.

In 1840, the year the Obstetric Division of the Vienna General Hospital had been divided into its two clinics, a German pathologist, Jakob Henle, had formulated a theory in which he supposed that disease could be caused by 'miasms', or polluted air, and that 'contagia' were miasms that had developed in the human body. There was nothing new in the idea that bad odours carry disease, as readers of the Bible or Shakespeare well know. Henle's originality lay in the fact that he supposed contagia to be organic, living things which survived on the human body as parasites do.

Henle's guess – it was no more – was a significant contribution to the germ theory of disease, and it was available to Semmelweis when he began his work on puerperal fever. But like the rest, it was only a theory. Henle had done nothing to confirm it and had no technique by which he could do so. There is no evidence that Semmelweis knew of it. And even if he had, there were too many other theories available to him to assume that it might have influenced him unduly. Semmelweis was an obsessive, pragmatic Hungarian. As he strode round the wards to see what he could see, no existing theory of that time could tell him what practical measures to take to eliminate the disease causing one bed in ten to be emptied of a dead woman.

The patients and midwives in the hospital would talk freely to the fat, rather humourless foreigner with the forthright manner. When some whispered that the cause of death in the first division's delivery ward was due simply to the fear that women had of being there, he took the suggestion with absolute seriousness; he did not reject the possible psychological origin of illness. Besides, he had already had the affecting experience of women crouching at his feet, clinging to his trousers, begging not to be sent to the first division. No matter how naïve his approach, the control group of patients at his disposal made it a relatively simple matter to test, falsify and discard what was of no use.

He went to some trouble to dispose of the 'fear' theory. He had noticed during the time he spent in the five wards in his clinic that the daily event which more than any other evoked a reaction of dread and worry among the women lying in their beds was the sight of a robed priest and the sound of his attendant's bell as they walked through the wards to reach the sick-room to give the last sacrament to a dying patient. Semmelweis also noticed that the priest's route from the chapel to the sick-room of the midwives' clinic was not through the wards. A quiet word from Semmelweis soon had the man walking to the sick-room by a roundabout route and without his bell. The result had no effect whatever on the deathrate in the first clinic. The simplicity of the technique, however, encouraged Semmelweis to look for more differences in procedure between the teaching ward and the midwives' ward, whose influence he could easily assess.

He rejected ideas about some of the differences without bothering to apply them in a comparative study with his control group. For example, there was a suggestion that childbed fever in the first division was linked to the low social class of the patients by the sense of shame felt by poor women when they were being examined by medical students. Semmelweis argued, not unreasonably, that the high-society patients who were lucky enough to be attended by a doctor in their own homes ought to suffer an even greater wound to their modesty at being handled by a strange man; yet the incidence of puerperal fever in these home deliveries was far less than in hospital.

Another suggestion was that the examination of patients by the midwives was an altogether gentler process than that carried out by the students in their first division teaching rounds. But how small, said Semmelweis, is the likely damage of a student's finger in the long vagina of a pregnant woman compared with the injury inflicted by a baby on

31

its birth passage? His supposition was correct, but had he used his control group as a verification it might have led him to some related important conclusions.

Semmelweis compared as many of the conditions in the two clinics as appeared relevant. Ventilation was identical, there was no perceptible difference in the standard of cleanliness, the laundry was washed in the same way, food was supplied by the same caterer. There *was* no obvious solution. However, Semmelweis's emotional involvement with the problem confirmed its existence at every turn. There was a poor and worsening life-expectancy of women lying in the teaching clinic. But there was one group of women which was the rule's exception.

The General hospital was a charitable institution and the city was vast. Many poor women arrived at the gates having failed to reach the place in time by foot, carrying their babies in their arms. They had given birth in shop doorways, under archways, in horse-drawn cabs, or anywhere they could find shelter. Yet the mortality rate from puerperal fever among these women was low and, Semmelweis found, roughly the same in both divisions. It was a consolation for poverty.

The other embarrassing statistic which Semmelweis extracted was that under the regime of the previous professor of midwifery, mortality had been extremely low. Under Professor Klein, it was extremely high. A tactful junior doctor would not have bandied about the information carelessly. But Semmelweis's uncontrollable volatility and emotion were incompatible with tact. Klein, his regime and methods, were firmly implicated with the cause of the high childbed fever death rate.

Klein had the reputation of being a dull, unadventurous and insecure man. Semmelweis was humourless too, but also manically depressive and volatile by turns. It seems, in retrospect, that however Semmelweis had presented the results of his findings to Klein, the characters of the two men were such that Klein's instinct to territorial protection could not fail to surface. When one of Semmelweis's more senior friends at the hospital, Skoda, a diagnostician with a growing European reputation, proposed that a commission of inquiry should meet to investigate Semmelweis's statistical results, Klein felt bound to take action. He invoked the protection of the Minister of Education and the commission was never allowed to meet.

But when in 1846 the numbers of deaths in the first division from puerperal fever reached more than four times that in the second, a commission did meet and made a recommendation. It was that the most likely cause of the disease was injury to the genital organs inflicted by rough examination by students during their course of instruction. More particularly, it was concluded that the injuries were caused by foreign students whose numbers should be reduced. Xenophobia had been introduced into medical judgement.

Semmelweis, the foreigner, watched his students reduced from 42 to 20, with foreigners almost completely excluded. He also watched his statistics. Mortality in his wards at first dropped, and then rose to frightening heights. By April 1847 it had reached 18 per cent.

During this period Semmelweis had one of the deep fits of depression which he experienced at different times throughout his life. His needed little to trigger its peaks and troughs. On this occasion, however, there was a ready excuse if one was needed. Throughout the clinic Semmelweis could sense a deep feeling of contempt among his patients and even among the domestic staff for the medical staff. The laymen were convinced that the medics were themselves in some way responsible for the appalling deathrate. Semmelweis himself stood among the accused.

He looked for obvious reasons which might support the patients' accusations, but there were none. In desperation, when he noticed a difference in the method of delivery used in the two divisions, he ordered that his own delivery ward should adopt a new method. He wrote:

Like a drowning man clutching at a straw, I gave up the dorsal position in labour, which was customary in the first clinic, and because the lateral position was that adopted in the second clinic. I do not believe that the dorsal position was so disadvantageous compared with the lateral position as to cause the higher mortality.

He was right. It made no difference whether women were delivered on their backs or on their sides. In the first division clinic they died at an unchanging, appalling rate.

So far Semmelweis had succeeded only in stirring up antipathy towards himself, by emphasising a problem which reflected the shortcomings of his superior. He had challenged orthodoxy, but had not replaced its techniques with anything preferable. It cannot have come as a complete surprise to Semmelweis when, in October 1846, Klein declined to renew his assistantship, but offered a provisional appointment instead. Semmelweis, touchy and haughty, was offended. He brooded over his future, left Vienna, and half-heartedly took up a study of English with a view to going to Dublin to investigate puerperal fever there. But by February of 1847 his old post was again vacant and he was reappointed.

Back in Vienna in a refreshed frame of mind, he had been walking the wards for only a few hours when the news reached him that one of the professors he admired most in the Medical School's faculty, Kolletschka, was dead. While carrying out a post-mortem a pupil's knife had slipped and pierced Kolletschka's finger. He had died his painful death within a few days of an infection setting in. The symptoms were described to Semmelweis, who reacted in his typical hypomanic fashion:

In the excited condition in which I then was, it rushed into my mind with irresistible clearness that the disease from which Kolletschka had died was identical with that from which I had seen so many hundreds of lying-in women die . . . Day and night the vision of Kolletschka's malady haunted me, and with ever increasing conviction I recognised the identity of the disease.

In fact he only partly understood Kolletschka's death. Semmelweis's theory was that the knife had infected him with 'cadaveric particles': minute decayed pieces of flesh from the dead body. At this stage the important question, what are 'cadaveric particles'?, did not occur to Semmelweis. He was simply concerned with preventing them reaching the open wounds of the genital organs of a woman after childbirth.

It was not now difficult for him to think up a reason how they got there. Obstetric students in the first clinic frequently moved from autopsies being carried out in the post-mortem room which joined directly onto the clinic's delivery ward, to the examination of women in labour. They went through an apologetic hand-washing procedure before moving from death to birth, but it was well known that the stench of a dead body clung to the hands for hours after washing with ordinary soap and water. The particles, Semmelweis realised, were carried to a women's sexual organs by students' and teachers' dirty fingers.

Semmelweis now insisted that his students wash their hands in chlorine water before they began any examination in the first clinic. He had invented an antiseptic procedure.

The first students to wash their hands in chlorine water did so in May 1874. The solution had an acid smell, was unpleasant to work with, and was expensive. Semmelweis soon substituted chlorinated lime which he rightly believed would have the same effect as chlorine water.

The result was a sensational success. In the seven remaining months of 1847 Semmelweis reduced mortality to 3 per cent. The 1846 level had been 11 per cent. In the midwives' clinic mortality was 2·7 per cent. In 1848, for the first time in its history, mortality in the teaching clinic fell below that in the midwives'.

Semmelweis was elated and, on the peak of one of his hypomanic phases, rigorously insisted that every student, teacher and medical visitor who wanted to carry out an examination of any sort should wash his hands in chlorinated lime on entering the labour ward, then use soap and water between examinations.

In the short term, particularly in the eyes of Semmelweis's reactionary seniors, there was no reason to assume that this newfangled and messy procedure was other than another piece of mumbo-jumbo, coinciding with a lull in the virulence of the disease. There was every indication that this was so, shortly after a pregnant woman who was suffering from a cancer of the cervix (the neck of the womb) was brought to the labour ward in October 1847. She was put in bed number 1. The hand-washing ritual was followed by all who entered the ward, under the priest-like supervision of the obsessive Hungarian doctor. Nevertheless, in spite of these precautions, in deeply-felt horror, Semmelweis watched eleven of the twelve women in the ward die of puerperal fever.

He was convinced that the position of the woman's bed and the fact that she was first in order of examination was critical in the spread of the disease. He was also still convinced that an examining finger had carried it. If this was so, then it must be the case that the infective particles, whatever they were, could come from the living matter of the human body; in this case a cancerous woman. The dead bodies of the post-mortem room were evidently not the only source. To prevent this sort of carnage happening again, Semmelweis insisted on the disinfecting chlorinated lime-wash being used between patients. Soap and water would not do.

But within a month another tragedy was played out in the same labour ward. Semmelweis needed to experience it in order to give a fine focus to his theory. A second sick woman was admitted into the labour ward. She had a bad ulcer in her left knee joint. The sore was open and running and its smell filled the delivery room. The woman's genital organs, however, were in no way infected and there seemed no reason why she should not have a safe delivery. Yet again, however, in spite of every care in the hand-washing procedure, the ward was decimated.

Semmelweis's conclusion was that, 'The air of the labour room, loaded with the putrid matter, found its way into the gaping genitals just at the completion of labour, and onward into the cavity of the uterus where the putrid matter was absorbed, and puerperal fever was the consequence'. From then on he kept any badly infected cases out of his labour ward.

The statistics of mortality for the months which followed were unambiguous, and in the year following Semmelweis's experiences with the two infected women, mortality fell to 1·3 per cent.

Unlike Henle, Semmelweis did not really care what the 'decomposed animal-organic matter' was which was carrying the disease through his wards. But he deeply cared about putting his theory to the test and he was able to bring extraordinarily effective methods of disease-prevention into operation. In his depressed periods he also carried a deep sense of guilt at his long ignorance: 'God only knows the number of women I have consigned prematurely to the grave.' Besides this he had the consummate knack of making others feel guilty: his professor in particular. The statistics which Semmelweis dredged up from the past showed unmistakably how puerperal fever mortality had increased dramatically in the year Klein took charge of the department. Klein's predecessor had taught midwifery on a model of the human body, Klein himself used dead bodies from the labour ward. The inference was clear. Students' fingers – and Klein's fingers – had carried childbed fever back into the labour ward. Klein was one of the woman-slaughterers.

This was only one of a series of factors which put Semmelweis in conflict with those who were able to make decisions on his academic future in Vienna. One of these decisions was to reject his application for promotion. But even before the first serious abrasions with his department's superiors, Semmelweis showed signs of a persecution complex somewhere beneath the surface of his thin skin. His response to the reluctance even to set up a simple investigation of his results was to take deep offence and to react irrationally. He refused an invitation to address the Vienna Medical Society on his work.

This could have been Semmelweis's opportunity to put his conclusions on early record. Instead, a few of the friends who were accustomed to the quirks in his personality took it on themselves to publicise his work. Semmelweis was more than fortunate in the quality of three of the young professors at the Medical School. Rokitanski, Hebra and Skoda, who all eventually won international reputations for themselves, immediately gave him their support. Skoda, at the Vienna Academy of Sciences in 1849, described Semmelweis's work as 'one of the most important discoveries in the domain of medicine'.

Skoda's opinion was no more than just, but Semmelweis took none of the steps to make sure that its importance was communicated. He reacted to Skoda and Hebra's good intentions with the ingratitude typical of his personality, blaming them for laying too much emphasis on contagion by 'cadaveric particles' and not on the rest of his theory. When Semmelweis did at last address the Vienna Medical Society in May 1850 with an account of the work he knew to be unorthodox and anti-establishment he, again typically, saw the criticisms of his work as personal attacks on himself.

When his second application for promotion was heard and granted, but without the full teaching privileges he ought to have been able to expect after six years' apprenticeship, Semmelweis broke with the Vienna Medical School. He stormed out of the city, back to his own country without even a word of goodbye to the friends who had tried to promote the ideas he himself has so inadequately promoted. Hurt by the conspicuous ingratitude, Skoda never spoke again to Semmelweis.

The abnormality in Semmelweis's character was unalterable, and so was his single-mindedness. Once back in Hungary he took the post of unpaid senior physician in the Obstetric Clinic of a Pesth hospital, reproducing the theory and practice in childbed fever preventive techniques. He called it his 'doctrine' and, still without publishing fully, he furthered its cause as enthusiastically as he had in Vienna, collecting just as much antagonism and reaction in the process as he had in Austria.

Just as Semmelweis's psychopathology caused his personal inadequacies, it was responsible for his prominent streak of scientific creativity. The creative energy he threw into his stream of work and activity is typical of many manic depressives. The pattern protected against the threat of depression and promised creative results from which could be won esteem and approval. But in Hungary Semmelweis was never to produce the originality of thought and approach which had marked his few years in Austria. Tragically too, he was now separated from the mainstream of European scientific thought into which, once injected, his ideas and his influence might have spread.

During these years of exile in his own country, Semmelweis became fatter, more florid and balder. A family life – at 38 he married a girl of 18 – might at first have given him a more stable pattern of existence, but he was still quick-tempered and needed to be soothed in his outbursts. And there were early unpredictable tragedies in his married life, which carried the mark of that irony which inevitably is attached to some part of the lives of all those who made contributions to the germ theory of disease. Semmelweis, who had spent so much of his life depressed by the deaths in his wards of mothers and children, saw his first child born a hydrocephalic. It was dead within forty-eight hours. A year later a second child was born, only to die within months, of peritonitis.

Semmelweis was accustomed to death, but these two in particular cannot have contributed to the stability of his personality. Over the years his friends watched his quirks turn into serious character defects.

For years he maintained that he loathed the idea of putting his work on paper, and the only account of his work was still that in the recorded minutes for 1850 of the Vienna Medical Society. When ten years after his last really original work in Vienna, he decided to take up his pen, words flowed from it in torrents. In 1857 he began to organise the statistics he had collected over thirteen years and set them down with his obstetrical observations and emotional experiences. It took him three years to write his *Aetiology, Conception and Prophylaxis of Child-bed Fever*. It was a rambling, repetitive, polemical and egotistical work, over-amply emphasising his misrepresentation and persecution; its 543 pages were loaded with such phrases as, 'Fate has selected me as the champion of the truth . . . a duty laid upon me which I cannot refuse to perform'.

The work had been completed in a state of high excitement, with Semmelweis furiously writing and rewriting chapters, then packing them off to the printer without correction. When *Aetiology* was finally published, it produced no reaction whatever.

Semmelweis's simple theory was that puerperal fever is caused by the transmission of organic particles to the open wound of a freshly delivered woman. Yet he failed to place his ideas in the common current of nineteenth-century scientific thought. And when, too late, he attempted to publish, his simple idea was so verbosely expressed as to defy attention. Semmelweis had translated a whole sentence in the vast scientific hieroglyph, but had failed to put it in sequence. His inability to communicate unmistakably what he had done condemned his work to sterility and thereby lost him his claim to scientific creativity.

When the book which had taken three years to complete failed to produce the response he craved from the medical press, Semmelweis clung desperately to his pen as his saviour. Now he began to dash off long emotional open letters either to those he felt might have influence, or to those who opposed him. Their contents were bitter, irrational and accusatory. One professor of midwifery in Vienna took the brunt of his attack: 'In this period of ten years at least 1924 patients lost their lives from avoidable infection . . . In this massacre you, Herr Professor, have participated'.

By the middle of 1865 Semmelweis's behaviour had become so eccentric in public and private that it was clear he was suffering from mental illness. His wife consulted one of his colleagues from the Medical Society where, even in lectures, Semmelweis was becoming an embarrassment. A consortium of his medical colleagues gathered to recommend treatment. What they proposed should be done to him – blood letting and cold-water dousing – was the sort of irrational treatment his methods should now have been helping to eliminate from medical procedures.

Eventually he was taken back to Vienna. There one of the few remaining friends from his early years in the city, Ferdinand Hebra, invented some reason to persuade the prematurely aged Semmelweis to enter the gates of a mental institution.

In the asylum a medical examination showed that Semmelweis had an injury to a finger on his right hand. It was believed to have been accidentally inflicted during his last obstetric operation. The wound appeared gangrenous. The last irony was particularly bitter. He died, as Kolletschka had done, of a puerperal-like infection, and in the same way that had made Semmelweis himself realise the mechanism of the infection. He was taken back to the Vienna General Hospital for the first time in fifteen years – in a coffin. There an autopsy showed extensive organic brain damage.

CHAPTER FOUR

One page in the long and complicated history of science is liberally sprinkled with the names of men obsessed with the value of a single idea: men who have devoted the best part of their lives to getting it accepted. Sometimes they succeeded and sometimes they had nothing better than a footnote at the bottom of the page to show for their efforts. Much more rarely a man emerges, often one whose character and skills are elusive, but who is clearly extraordinary in that he has the ability to produce scientific creativity almost at will. Louis Pasteur was such a man.

Pasteur was born within four years of Semmelweis: their first significant work was produced at about the same time and some of the research done by Pasteur as a relatively young man, had it been known and understood by Semmelweis, could have turned the theory arrived at by the Hungarian in his crowded labour ward into medical practice of vast significance. Instead, and Semmelweis cannot escape blame for this, his work stagnated while people such as Pasteur, in ognorance of its existence, bypassed it.

Semmelweis's contemporary influence on his fellow man's well-being was negligible. The same cannot be said of Pasteur. At Pasteur's birth-date in 1822 life expectancy at birth was less than 40. When he died, in 1895, that figure had been increased by many, many years and would continue to rise spectacularly; in another fifty years it would almost double. Much of that increase is directly attributable to the consistency of the work of Pasteur and of those who built on the foundations he laid. If greatness exists in science, Louis Pasteur was indisputably great.

Attempting to analyse genius is not easy. Yet, quite early in his life, some of his contemporaries easily recognised genius in Pasteur. He was a small man who was capable of inspiring devotion bordering on worship in his family and friends. One friend who came under the spell of the god-like little man, looking into his grey-green eyes, described him as having 'the sparkle of a Ceylon gem' in which 'the flame of enthusiasm shone'. His young wife, the daughter of the Rector of Strasbourg Academy, needed no encouragement whatever to dedicate utterly her existence to Louis' scientific ends. She would spend long evening hours writing to his dictation, willingly accepting the fact that what happened in the laboratory that day was of prime importance in their lives together. But, just as easily as he inspired devotion, Pasteur's aggressive manners were capable of rousing antagonism and bitterness in the breasts of many men who worked at his own intellectual level. One of his strongest motivations was an apparently uncontrollable desire to demonstrate his intellectual superiority in as nihilistic a fashion as possible.

Some of Pasteur's motivations seem extraordinary in one working in the field of science which, tradition says, prospers best in conditions of purity of thought. Pasteur's case shows very well how less attractive is reality. The impetus to creativity in science can be as ordinary or banal as the impetus in less idealised spheres of mankind's operation.

There is no question that, of all unlikely motives, nationalism – that obsession with the destiny of France which has been the deep concern of so many Frenchmen – was a powerful force in Pasteur's life and thoughts. His father, a tanner who had served as a non-commissioned officer in the Peninsular War, and for whom Napoleon was a demigod, fostered his own attitudes in his son. Jean Joseph Pasteur had returned from the wars with his sword, and a Légion d'honneur which he wore in such a way that it could be seen without much difficulty at fifty yards. Louis continued his father's worship of the Emperor to an almost inconceivable degree. When his scientific fame put him on nodding terms with Napoleon III, he was to write to his own son, 'I have just had the honour of having been invited by the Emperor to spend a week with him. Such are the rewards of industry and good behaviour. So you too must work hard, to be rewarded, God willing, in the same way.'

The sycophantic intensity of Louis Pasteur's feelings for 'Napoleon the Little' were only exceeded by the passion of his feelings for his country. During the Franco-Prussian War he was to write:

I want to see France resisting to the last man and the last defence-work, I want to see the War prolonged into the depths of winter, so that, with the elements rallying to our side, all those vandals confronting us shall perish of cold and hunger and disease. All my work, to my dying day, will bear as an inscription, 'Hatred towards Prussia! Revenge! Revenge!'

And a few years later he was prepared to link France's rebirth with, of all things, science: 'The salvation of France has been encompassed by her superiority in the sciences and nothing else.'

Some of Pasteur's work was solely directed towards doing good for his country. When Napoleon III called, as he did when France's wine industry was in a perilous state, Pasteur answered, with a will. But there were other strong motives operating on this complex man. Personal ambition was vastly important. Pasteur was a conventional man who saw adherence to conventional behaviour as his duty as a citizen of France. He also accepted the idea that intellectual achievement in science was one of the noblest aims to which a man could aspire. He was not a modest man and he had no doubts about the ultimate achievement of his ambitions. He would spend days at a stretch in his laboratory, complaining that the nights were too long in keeping him from his work. He told a friend, 'I am often scolded by Mme Pasteur, but I console her by telling her that I shall lead her to fame'. This was before he was 30; he was not being flippant. He told his wife that the work he was doing would bracket him with Newton or Galileo.

The depth of the very conventional motives which spurred Louis Pasteur on is extraordinary. But the nature of his work was in no way conventional. It was itself a mosaic: a series of beautiful experiments which he made into a self-contained pattern as his massive contribution to nineteenth-century science. It began with a problem of pure chemistry but it ended with an understanding of human disease as it has never been understood. Pasteur forged the link between chemistry and medicine.

The boy from humble origins, son of the sergeant-major, Pasteur reached his career in chemistry as a young man through the École Normale in Paris, where entry was on a competitive basis. His first research work, which he began at the age of 23, had the appearance of being the choice of a very conventional young man indeed. There are fashions in science just as there are fashions in hats, or games, or attitudes. This is as true today as it was in Pasteur's day. In chemistry in Paris in the eighteen-forties, and for many years afterwards, there was a fashion for the study of crystals and, true to form, Pasteur adopted it. But here, as with all his work in science, conformity ends. Pasteur's approach was always the nonconformist, but strictly rational application of the method he had acquired in his years at the École Normale. He had a passionate belief in what he called 'this marvellous experimental method, of which one can say, in truth, not that it is sufficient for every purpose, but that it rarely leads astray'.

The problem on which Pasteur chose to carry out his research arose from a remark of the German chemist, Mitscherlich. Mitscherlich had come to the conclusion that there was something odd about crystals of tartaric acid and racemic acid. He believed that they were identical in every way – same composition, same behaviour in chemical reactions, same crystal shape, and so on – except that a solution of the tartrate rotated the angle of polarised light, while a solution of the racemate did not.

Pasteur reasoned they could not be identical. His approach to the problem was clear, and the pattern he followed in investigating it can be found in all his later work. First, he identified precisely what his problem was; he then sat down and taught himself as much as he could about the past history of the subject; he then guessed what the answer was

43

likely to be. There is no question that Pasteur always worked from
theory to practice, and not the other way round. He himself stated quite
clearly how he operated: 'Without theory, practice is only routine
governed by the force of habit. Only theory can breed and develop the
spirit of invention'. The common idea of the scientist sitting at his
laboratory bench, observing nature, collecting data, then inventing a
theory which fits the observed facts, in no way applies to the work of
Pasteur. Having made his guess – his theory – he then set out to devise
an experiment to test it.

In the case of the racemate and tartrate, Pasteur's guess was that it
should be possible to see a difference in the shape of their crystals. With
great skill and patience Pasteur prepared his crystals and looked at them
under a microscope. What he saw had been missed by everybody else
who had looked at similar crystals. Each one had a tiny face (or facet).
The tartrate had this facet orientated in the same direction on every one
of its crystals; they were, so to speak, right-handed. In the racemate,
however, Pasteur found two sorts of crystals: in each case, the facet
sloped in a different direction: it was a mixture of right-handed and left-
handed crystals. The distinction was so clear that Pasteur could pains-
takingly separate the two sorts by hand.

He now had two sorts of racemate crystals: one which looked different
from the tartrate, and one which looked identical. And it was identical.
He found that it rotated the angle of polarised light in the same direction, and by the same amount. The other racemate rotated the light by
exactly the same amount in the opposite direction. He now knew why the
racemate was optically inactive: it was a mixture of two different substances, each of which, when exposed to polarised light, cancelled the
effect of the other.

Pasteur knew instantly that he had made a great discovery and that
he had passed the first stage in the route along which his personal
ambition drove him. He rushed out of his laboratory and into the corridor
to grab the first man he could see. He happened to be one of the laboratory assistants. It mattered not in the least to Pasteur. He embraced him
and dragged him across to the Luxembourg Gardens and poured the
results of his first creative efforts into the captive and probably only
half-comprehending pair of ears.

This extraordinarily clever piece of research was the solid beginning
on which Pasteur could establish the academic career he had set his
heart on. It led him to his teaching appointment in Strasbourg, where
he found his wife who was prepared to devote her life to him and to his
work, and then to Lille as Dean of the Faculty of Science. It was the

conventional progress he wanted: the first fruits of the unconventional and original mind. He took up new problems, but he continued his interest in crystals, all the while relating his new fields to his past discoveries to make a beautifully logical progression of his work.

The new fields were soon to have him puzzling over the nature of life itself. Life, as it was understood in the 1850s, seemed far removed to most chemists from the inanimate crystals with which Pasteur was working. But Pasteur had noticed an unlikely relationship. He observed in his laboratory one day that a living thing – a mould which happened to have grown on some calcium racemate – turned the racemate from being an optically inactive substance into an active one. What was more, he found that the effect became more pronounced with time. He showed that the mould was destroying that part of the racemate which rotated plane polarised light to the right. A living thing preferred a substance with a particular shape and rejected the other constituent of the racemate. He then swept on to the brilliant theory that optically active substances derive from living things. Pasteur wrote to a friend, 'I am on the verge of mysteries, and the veil which covers them is getting thinner and thinner'. It was true. He was near to pulling a cover from some hidden chapter of science. But such was the state of understanding of the processes of life that only he and few others had any inkling of what a complicated tale there was to tell.

In the summer of 1856 Pasteur had a visitor from a Lille industrialist who had a simple problem of profitability. Bigo, the industrialist, whose profits depended on the production of alcohol from fermented beet juice was getting a contaminated product. So were other alcohol manufacturers in the district. Could Pasteur, now known to be a skilled chemist, help?

This was the first time that Pasteur was asked to apply his science for money. The acquisition of cash was not a factor which motivated him, though his frequent lack of it often worried him. But he never made any pretence of being able to separate science from the applications of science, and many of his discoveries stemmed from requests which had profit as their sole bourgeois motive: profit for an alcohol manufacturer, a brewer or, more grandly, for the French nation. But in any case, alcohol was high on the nation's list of priorities.

Pasteur went to Bigo to look at the vats of fermenting juices. The manufacturer had defined his problem plainly. Pasteur, as he always did once the problem had been identified, had already familiarised himself with the existing background knowledge which might prove useful. He knew, for example, that one man at least believed that the yeast involved in fermentation was a living organism. Charles Cagniard de la Tour twenty years earlier had looked down his microscope to see tiny particles sprouting little buds in an unmistakably lifelike fashion. However, the orthodox view of chemistry at that time was that fermentation was a purely chemical process. The great German, Liebig, and the equally great Swede, Berzelius, both with world-wide and almost God-like reputations, taught that the complicated substances present in fermentation simply acted as catalysts, helping the reaction along, but remaining themselves unaffected.

Pasteur took samples from Bigo's vats back to his laboratory and looked at them under his microscope. In the samples from the healthy vats – those giving Bigo his healthy profit – Pasteur could see yeast cells sprouting little buds. But in samples from Bigo's sick vats, the ones failing to produce alcohol, Pasteur could see, not globules of growing yeast, but shimmering little rod-shaped organisms.

Pasteur studied many aspects of the problem. Then he formed his theory. It was that the yeast involved in alcoholic fermentation must indeed be a living organism, that it fed on beet juice, and that alcohol was the end-product of its metabolic processes. Different organisms, he guessed, produce different end-products. Those he had seen in the sick vats as a grey substance, and under his microscope as tiny vibrating rods, and which prevented yeast from growing, had lactic acid as an end-product. The same variety of organism could be seen in sour milk.

In that heyday of chemistry the suggestion that a chemical, drinkable product like alcohol was the waste from some small creature's stomach, or whatever served it for a stomach, was a terrible heresy. Today it is difficult to recognise just how contrary to established thought was this suggestion of Pasteur's. Establishment chemists such as Liebig and Wöhler would have none of it. Already they had lampooned the earliest germ theories with parodies describing minute creatures that ate sugar and deposited alcohol through the anus and carbon dioxide through the genitals. Pasteur was meat for more ridicule.

But Pasteur's method embraced a brilliant ability to be able to test his apparently wild theories. He showed that sugar *never* underwent alcoholic fermentation unless accompanying it were living globules of yeast. He also showed that the other organisms he had found in the grey material in Bigo's vats turned sugar into lactic acid. Pasteur's revolutionary (and to some, horrifying) conclusion was inescapable: minute living things can play vastly important roles in chemical processes. Micro-biology was born.

When the opportunity came for his academic career to move on,

Pasteur was ready. He was offered the post of director of administration and scientific studies at the École Normale in Paris. This, France's premier school, was at that time going through a bad patch in terms of scientific productivity. Pasteur had no intention of using this as a reason for rejecting the chance of going back to the capital. Paris was then one of Europe's liveliest scientific centres. Centre stage was where Pasteur needed to be: so that he could publish more easily, and so that he could be aware of what was being published: so that he could put into, and take out from the current of scientific knowledge. Pasteur was well aware of the dangers of having his work unnoticed, but such was his character and his powers to promote himself and his work that at no point of his career was there any chance of that happening. With his customary lack of modesty he took the École Normale job in order, he said, to restore the school to its former greatness. And, again customarily, his abilities were great enough to bear out his predictions.

Nevertheless, as an administrator of the school, Pasteur was a disaster. He did not have that sure touch in human relationships that he had when dealing with chemical processes. He was too harsh – a reactionary – when dealing with childish complaints about the food, or in handing out punishment for cigarette smoking. He was genuinely surprised and shocked to find that the disciplinarian's pattern he tried to impose on his students' lifestyles – that same pattern he had used to discipline their minds to cope with scientific problems – made him thoroughly unpopular.

There were no tangible compensations for the administrative work which Pasteur did not want, and did badly. No grand laboratories waited for him in Paris. The tradition then was that great things come from small beginnings. Many of France's nineteenth-century chemists and physicists were proud to have made their discoveries in dark little laboratories using apparatus that cost only a few pence. Pasteur accepted the tradition and carried out much of his work at the École Normale in a makeshift laboratory converted from a poky room in the attic. His incubator was a cupboard under a staircase. But, such was his experimental skill, the facilities had no effect on the quality of his work during the next few years.

Two years after the move to Paris, in 1859, Pasteur suffered the first of a series of tragedies that were deeply to affect his personal life. His nine-year-old daughter had been staying in the Arbois with her grandfather. There had been an outbreak of typhoid fever in the district and the small girl had been one of its victims. When death in most families was commonplace, this was the first time Pasteur felt involved. 'I cannot

47

keep my thoughts from my poor little girl,' Pasteur wrote to his father three months after she died.

A few weeks later he wrote to his friend, Chappuis, about his studies on fermentation, 'connected as they are with the impenetrable mystery of Life and Death'. Pasteur was now entirely clear as to the logical path up which his work was leading him.

He also told Chappuis, in his usual confident style, 'I am hoping to make a decisive step very soon by solving, without the least confusion, the celebrated question of spontaneous generation.' The 'celebrated question' was really a celebrated answer to how life came about. That it spontaneously generated itself had been an acceptable theory to many people since men first began to ask the question of each other. Maggots squirming out of rotten meat, and crocodiles pushing themselves up from the warm mud-banks of the river Nile, were, in the absence of evidence to the contrary, assumed to have originated their own lives in the surroundings from which they were seen emerging. Some quantitatively minded individuals even devised exact recipes for the creation of mice from sacking, flour, dust and a few other ingredients often to be seen lying around the floor of a baker's shop.

The naïvety of the concepts now seem laughable. Even so, it is as well to remember that the question of how life originated is today still far from settled.

In the seventeenth century Anton van Leeuwenhoek, that great figure in the history of technology, by his invention of the microscope, had made it possible to see a wholly unsuspected zoo of livings things in equally unsuspected places. What his lenses revealed was, in its day, an almost unbelievable sight which many – such were the places the creatures appeared to live – frankly disbelieved. Leeuwenhoek found the things in spittle, the scum on teeth, human skin, and his own semen. The 'animalcules' which he saw and drew, with legs, antennae and eyes, were as wonderful as they were weird and he took enormous relish in describing them. But his microscope gave no clue as to how they got where they were.

An eighteenth-century Italian priest, Lazzaro Spallanzani, set himself the aim of trying to disprove spontaneous generation by several well-conceived scientific experiments. He made a number of tubes, put into them various seeds in water, and heated his mixture. He now completely sealed some of the tubes, partly sealed others with wood and cloth, and left others open to the atmosphere. After some time, Spallanzani examined the material in his flasks in detail. The teeming life he saw down his microscope convinced him that the numbers of animalcules swimming around in each tube was dependent on the access the contents had had to the air. He believed that either the air carried the small organisms into the flask, or caused those already there to increase.

These were remarkable conclusions, considering the primitive laboratory equipment with which Spallanzani had to work. However, he scarcely dented the surface which protected the popularity of the theory of spontaneous generation. In the middle of the nineteenth century the doctrine was still blooming in the hands of several reputable French scientists. One of them, Félix Pouchet, director of the Natural History Museum of Rouen, proclaimed to no less a body than the Académie des Sciences that, 'animals and plants could be generated in a medium absolutely free from atmospheric air'.

It was a situation tailored for confrontation with the character of Louis Pasteur. Pasteur loved to make a battlefield of his science and here was an ideally vulnerable enemy. Pouchet had disturbed the latent vicious intellectual aggression which occasionally powerfully influenced Pasteur's actions. When he was offered the chance of showing the superiority of his techniques over those of his contemporaries, he seized it. If he could display that superiority publicly, so much the better.

The experiments which Pasteur devised to annihilate Pouchet were beautifully simple, and were the more devastating for their simplicity. The best known is that in which he put a fermentable solution in a balloon-shaped flask fitted with a swan-neck. Pasteur then heated the liquid for long periods in order to destroy any life existing in it. Next, he allowed the liquids to stand undisturbed, some of them for many months. The liquids remained quite clear. Pasteur's microscope showed that they contained no living matter and that no fermentation had taken place. If he shook the flasks, however, or if he broke off their swan-necks, then the liquids quickly clouded. The experiment confirmed Pasteur's belief that the living matter in the air of the laboratory was what caused fermentation; any germs which might have floated into the liquid as it cooled and drew air into the flask were trapped in the S of the swan-neck.

Pouchet and his fellow partisans of spontaneous generation were tenacious opponents. They argued that if air contained enough germs to cause almost any fermentable material to ferment, as Pasteur suggested it did, then 'A crowd of them would produce a thick mist as dense as iron'. It would not be possible even to walk through the air.

Pasteur's answer was to suggest that the air was not equally thick with germs. To prove his point he devised one of his most picturesque experiments. He set off with a party of assistants and a small baggage train of equipment for the Jura mountains. He carried with him seventy-three sterile, hermetically sealed flasks. Not far from his father's old tannery, though well away from any habitation or animals, twenty of these flasks were carefully opened. Eight became cloudy, showing that germs had entered with the air.

Pasteur's belief was that the higher he and his team climbed, the less likely was the air to contain living organic matter. From Salins they pushed on to Mount Poupet, 850 metres above sea-level. Holding the flasks high so that no living matter could enter from his clothes, Pasteur carefully snipped off the ends of his flasks, then resealed them with a pure alcohol flame. Five of the flasks became cloudy.

From Mount Poupet, Pasteur headed for Chamonix. From there, with an alpine guide, he trecked on to the Mer de Glace: ever upwards bearing his strange flasks. The simple experiment, carried out in the cold and wind of mountain conditions, sometimes with sunlight so brightly reflected from the ice that it was impossible to see the jet of flame, required great patience. In the final series of twenty flasks, only one was altered when it was opened to the atmosphere.

Pouchet, however, was neither an inactive nor an unworthy opponent. At the same time as Pasteur was battling with the Alps, he was struggling up Mount Etna – and reaching entirely opposite conclusions. He maintained that, no matter how pure the air, 'taken where you like', it eventually spontaneously generated legions of germs. Whereas Pouchet's technique of collecting air never improved, his physical courage soared and he bravely took on Pasteur on his own ground and in worse conditions. In order to prove his point he survived freezing nights on the bare Alps, and treacherous conditions which almost cost the life of one of his co-workers.

Pouchet withstood Pasteur's exhausting logic for three years before perishing beneath it. Pasteur demonstrated his superiority in a lecture at the Sorbonne on 7 April 1864. Pasteur was a skilled showman. It was one of the chief reasons why his work never passed unnoticed. To be able to humiliate an opponent intellectually in front of an audience doubled the satisfaction he got from his work. That evening he had a gallery of talent and glitter in front of him. Alexandre Dumas Sr, George Sand and Princess Mathilde were some of those who had come to watch Pasteur demonstrate his skills.

Pasteur's use of language was as beautifully precise as his experiments. He showed his swan-neck flasks to his audience, demonstrated his material, and explained what he had proved, drawing every ounce

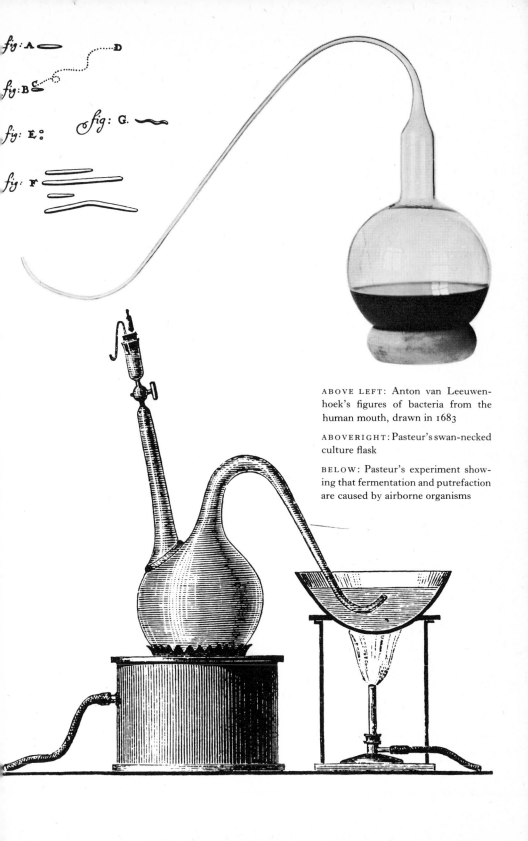

ABOVE LEFT: Anton van Leeuwenhoek's figures of bacteria from the
human mouth, drawn in 1683

ABOVE RIGHT: Pasteur's swan-necked
culture flask

BELOW: Pasteur's experiment showing that fermentation and putrefaction
are caused by airborne organisms

of drama from the magnificent situation he had devised for himself. He concluded:

And, therefore, gentlemen, I could point to that liquid and say to you, I have taken my drop of water from the immensity of creation, and I have taken it full of the elements appropriate to the development of inferior beings. And I wait, I watch, I question it, begging it to recommence for me the beautiful spectacle of the first creation. But it is dumb, dumb since these experiments were begun several years ago; it is dumb because I have kept it from the only thing man cannot produce, from the germs which float in the air, from Life, for Life is a germ and a germ is Life. Never will the doctrine of spontaneous generation recover from the mortal blow of this simple experiment.

He had rubbed Pouchet's nose in the settling dust; now he stamped on it by pointing out to his audience how ill-conducted the experiments of his opponent must have been: experiments ruined by errors which Pouchet 'either did not perceive or did not know how to avoid'.

At the same time that Pasteur had been applying himself to spontaneous generation, he had also been involved with what were, on the surface, less philosophical and more bluntly applied aspects of science. Pasteur had by now become the nineteenth century's first technological trouble shooter. French industrialists recognised that they had invaluable allies in the little bearded professor and his microscope. He descended one day on Orleans, the great vinegar manufacturing centre, took from the casks the living agent responsible for vinegar production, showed it to the manufacturers under the microscope and told them how to control its behaviour in the interests of maximum profits.

And when at last in 1863, Napoleon III himself called on Pasteur's help, the problem was of a large dimension. The wine industry, that bedrock of French economy, whose product was as much a national symbol as the very language itself, was in a desperate state. Vintners in many parts of the country were producing large quantities of diseased wine. Even at that time 2,000,000 hectares of French soil were under vine, and the average value of the product was 500 million francs. The figures involved explain the concern at highest governmental, and even Imperial level.

Pasteur, when he came to stand with his microscope alongside the vintners, was able to stagger their beliefs: in particular, the faith they kept in their palates as the only good judge of wine. By looking through his eyepiece at samples of the liquids, observing what he called the 'parasites' present, he was able to announce whether the wine was 'turned' or not, and how it would taste as a simple deduction from the living matter he could see in it.

It was clear to Pasteur that after wine had been fermented, potentially harmful organisms remained in it which ought to be dispensed with. He had a simple solution to the vintners' problem. Heat the wine gently to kill off the organisms. The wine could then be aged without it going sour. Horrified as they were by this heretical suggestion, the connoisseurs, in the interests of their profits, gave the method a trial. *Pasteurisation* was born.

The first practical application of Pasteur's work to animal disease came when he was asked by the celebrated chemist, Jean-Baptiste Dumas, also the Minister of Agriculture of the day, to investigate the cause of pebrine, a disease then troubling the silkworm breeders of Southern France to the tune of millions of francs. In ten years the French silk industry had been reduced to producing one seventh of its customary output by the blight.

Pasteur knew nothing at this stage, either about silk or silkworms. But the problem that silkworms could no longer make silk because they were dying of some unknown cause, was one he could approach with the simple methods he had already used and the experience he had gained. It would take six years of work to solve the silk growers' problem, but within a short time of his arrival on the scene Pasteur was able to demonstrate a most important conclusion. The disease was contagious, and the contagion, which could be spread in the air, or by those tending the worms carrying infected matter on their clothes and hands, was due to a living organism. It was not bad work for a man who, until this inquiry, had scarcely so much as seen a silkworm. It was also a vital step taking Pasteur further forward towards relating germs to disease as a whole.

During this period, and within weeks of one another, two incidents forcibly turned his attention to human disease and the vulnerability of the human body. First his father, whose attitudes had so dominated his own, died. Then his two-year-old daughter, Camille, was shown to have an inoperable tumour, and quickly succumbed. Pasteur carried the small coffin to Arbois to put the child next to her sister and his father.

Only a few weeks after the death of his youngest daughter, Pasteur began the first of his attempts to understand the nature of human disease. Late in the summer of 1865 cholera hit Marseilles, then Paris. By October it was sweeping up 200 victims a day. It was the opportunity Pasteur needed to try to prove what was still no more than a hunch, that a single organism could cause a single disease in a human being.

The opportunity provided by the cholera onslaught, however, also

53

carried a serious risk to the investigator, who was very vulnerable if the theory propounded by Pasteur made sense. Therefore the three distinguished men who squatted beside one of the ventilators leading from a cholera ward of the Lariboisière hospital during the 1865 outbreak, were also brave men. They were Pasteur, Claude Bernard and Sainte, Claire Deville. They had adapted a glass tube to the exit of the ventilator and surrounded it with a refrigerating mixture in order to try to condense whatever was present in the air of the ward. The three survived, but their experiment gave them no clue as to the specific origin of cholera. All Pasteur could see down his microscope, looking at what he had collected, was a mass of different floating particles including a variety of organisms. The germ – or microbe as it was later to be called – which Pasteur was looking for, if it existed, could not be singled out. Nor could he discover it in the blood of cholera victims. This was his first significant experimental failure.

Pasteur's fame soon took him to spend a week in the worldly apartments of that symbol of authority which he most venerated: the Emperor himself, Napoleon III. At the Château de Compiègne Pasteur was irrationally and unaccountably bowled over by the magnificence of the monarch, and the flummery and luxury which surrounded him. The professionally rational and critical Pasteur displayed a breathtakingly uncritical view of the life-style of a man of little substance, and was more than flattered to be asked to explain in drawing-room language what his work was all about. He was pleased to communicate to the monarch the scientific ambition that had been developing in his mind since he had begun to study micro-organisms. Pasteur was able to explain how the creatures were responsible for the processes of putrefaction, and how he recognised that, in some way, they were involved in contagious diseases.

In Pasteur's case the irony of his helplessness in the face of disease was doubly sharp. He had some inkling of what the origins might be, but these presumptions told him nothing about how to begin to prevent a disease. In the spring of 1866 typhoid fever again struck in what was left of his close family. Again he was called too late to the sick-bed of one of his daughters. Again he had to take a coffin, this time that of his twelve-year-old child, Cécile, and bury it with the rest at Arbois.

Pasteur had explained to the world how microscopically small living things played an essential role in fermentation and in putrefaction, and he had already taken the first steps in explaining how a disease in silkworms was carried by the air. He was already convinced of the logic of the step relating germs to human disease; but this step involved that complicated organism, the human body, and was vast. So far there was no correlation whatever between any particular organism and human disease. In reality the enigma of diseases which spread themselves, and the vulnerability of the human body, were as little understood as they had been after centuries of so-called scientific medicine. The average age of death of the male was still not much more than 40.

At the age of 45, on the morning of 19 October 1868, Pasteur was taken ill. A fit of shivering forced him to lie down after lunch. When he insisted on walking to the Académie des Sciences in the afternoon, his wife was so alarmed at his condition that she persuaded one of his elderly colleagues to walk alongside him. That night Pasteur suffered a cerebral haemorrhage which gradually paralysed the whole of his left side. A leading doctor of the Académie de Médecine was fetched, who prescribed the application of sixteen leeches behind the ears. Pasteur himself was by this time too far gone to protest at this irrational and idiotic piece of medicine. However, the free flow of blood seemed to coincide with an improvement in the patient's condition. What was apparently the terminal illness of one of France's most exceptional scientific talents attracted the attention and real concern of the scientific community. But Pasteur had promoted his work, and himself, so well that interest in his death-bed spread far wider than the salons of science. Each morning a liveried footman in the service of the Emperor and Empress came for news of Pasteur's progress, which he carried away in a sealed envelope. Napoleon's Aide, General Favé, came to visit and brought with him a copy of the translation of an English book, Samuel Smiles's series of biographies of courageous lives, *Self-Help*. It was an attitude to life which, of all people, Pasteur least needed to be implored to adopt. In spite of the traumatic attack, in spite of the leeches, and in spite of the pessimism of the members of that medical profession of which he was not a member, the Imperial footman's visit decreased in frequency and urgency. Pasteur was recovering.

CHAPTER FIVE

Between 1844 and 1846, when the railway mania was sweeping Europe, the House of Commons authorised the construction of over 400 railways throughout Britain. All the grand clichés are true. The railway age altered the face of the continent and revolutionised trade and relations between nations. But there was another most important consequence. The steam technology that had sprung from the application of simple scientific principles fed on its origins. The railway had the most profound effect on the nature of the operation of science itself. Quite suddenly, laboratories in Berlin, London and Paris were no longer remote little islands of learning where men were simultaneously working on the same subject without knowing it, or really caring about what was happening elsewhere. Almost overnight most of the academic centres of Europe were within forty-eight hours of one another. Suddenly scientists were able to visit laboratories in distant capitals with not much more trouble than it took to get from Oxford to Cambridge. They formed new friendships, began to work abroad in far greater numbers, married one another's sisters – and, of course, communicated their work to one another.

At the same time as rapid railway postal communication became commonplace, the technology of high-speed printing was being perfected. Soon a publication from, say, a Paris laboratory was, give or take a few hours, as available to a research chemist in Heidelberg or Manchester as it was to one in Lille or Montpellier. The railway radically narrowed the time gap between scientific discovery and its application. As much as any other, this nineteenth-century piece of technology set the pace of competitive twentieth-century science.

In April 1856 a young Quaker surgeon finished a few weeks' honeymoon in the English Lakes with his plain, quiet but devoted new wife, and set off with her on a wedding tour of European medical schools. The distance Joseph Lister and his bride covered in their three months' trip would have been unthinkable a few years earlier in the pre-railway age. The Listers took in Brussels, Cologne, Milan, Pavia, Padua, and any number of other less known towns.

In Pisa Cathedral they sat through what Lister found a 'rather sickening ceremony', when an Archbishop in full dress chanted a prayer to the city's patron saint, asking him to prevent the recurrence of cholera, and the sickness of the vines. At this time animal and plant diseases were more securely linked in the Archbishop's mind than they were in Joseph Lister's.

In Vienna Lister visited the largest and most important medical school he had ever seen. There he was entertained by the man who was now Europe's leading pathologist. Rokitansky, once the committed supporter of Ignaz Semmelweis, had fulfilled his early potential.

Lister and his bride sat at Rokitansky's table, ate mountains of salted ham, goose, potatoes, beans, cheese and cucumber. The two men talked pathology and surgery; later Rokitansky spent hours proudly showing Lister his teaching wards and his museum. Yet, during Lister's stay of a fortnight, the name Semmelweis was never once mentioned, so signally had the Hungarian failed to communicate the wide implications of his work, even to his friends. Semmelweis had been forgotten.

Lister returned to Britain and his work in Edinburgh Royal Infirmary, uninfluenced by the discovery which might have been one of the most practically significant advances in the understanding of the nature of disease.

However, there was no possibility of Louis Pasteur's work evading Lister through any fault of Pasteur, so effectively had he communicated it. It got through to Lister even though, having acquired a busy professorship in surgery at Glasgow, he read relatively little of the increasing amounts of continental scientific literature. It was the professor of chemistry, Dr Thomas Anderson, who drew Lister's attention to Pasteur's writings. These writings were entirely straightforward and clear, and Lister was attracted by plainness and clarity. He immediately understood the implications of the germ theory for his own work. What Pasteur had to say about putrefaction was the making of Joseph Lister and, ever after, Lister knew it.

Lister was also a gentle man whose Quaker upbringing suited his character. The strongest expression his friends ever heard him use in his stammering speech was, 'It's an infamous shame'. Compassion for one's fellow men is a severely over-emphasised motive for the work of many who have made advances in medicine. Familiarity with medicine and its human subject-matter frequently dulls compassion. But Lister never lost his tenderness for his fellow human beings, nor his awe of the human body. The good he could do undoubtedly genuinely moved him. It was never, as with many, a retrospective motive. Though he made his reputation as a surgeon, all through his life a serious operation used to make him sick with anxiety before it began. 'To introduce an unskilled hand into such a piece of Divine mechanism as the human body,' he said, 'is a fearful responsibility.'

But in spite of seeing some spiritual force operating on the organism, he had, for his day, an unusually scientific approach to medicine which he undoubtedly acquired as a child. His father was a skilled amateur microscopist who, like Anton Leeuwenhoek, ground his own lenses. His invention of the achromatic microscope was one of the greatest contributions to nineteenth-century microscopy. Throughout his life, Lister wrote to his father of his scientific-medical progress, giving a continuous detailed record of his concern for human life, as well as for the human body.

Lister came to surgery in the early eighteen fifties when the first use of anaesthetic had taken some of the horrors out of the procedures that took place on the temporarily blood-soaked table over the permanently blood-stained floor. In 1826 the sensitive student of medicine, Charles Darwin, had watched two operations, one of them on a child. The screams, the struggles, the blood, were more than he could stand. 'I rushed away before they were completed,' he wrote, 'nor did I ever attend again, for hardly any inducement would have been strong enough to make me do so; this being long before the blessed days of chloroform.'

Most operations were amputations of limbs, since the rest of the body was believed to be outside the scope of surgery. The number of operations, not surprisingly considering a patient's chances of survival, was small. At one leading London teaching hospital, University College, the number of operations annually performed in the eighteen sixties was only about 200.

Mortality was frightful. Military surgery under operational conditions had the worst of all reputations. On some occasions there was a 90 per cent death rate. However, in the Crimean War, between 1853 and 1856, at a time when Joseph Lister was beginning his surgical career, when Guards regiments alone lost 2162 of Britain's most prized soldiers, 1713 of these died, neither from battle wounds nor the trauma of surgery, but from disease.

Disease, rather than surgery, killed those who submitted to it and, even after the introduction of anaesthesia, survived it. The infamous 'hospital diseases' – erysipelas, pyaemia, septicaemia and the horrifying black hospital gangrene – swept up victims more surely than surgeons themselves, and made surgical wards nightmares of smell, suffering and slaughter.

The understanding of the mechanism by which these and other diseases spread themselves was, in the middle of the nineteenth century, still almost nil. The only undisputed facts relating to prevention were that dirt and overcrowding increased the risk of the diseases, and these facts, used by Florence Nightingale in her battle with the War Office, led to the imposition of some order of cleanliness on the filthy hospital conditions of the British Army in India.

As a 19-year-old student, Joseph Lister had seen the first major operation to be performed with ether as an anaesthetic. A butler named Frederick Churchill, suffering from cancer of the thigh, had his leg removed. The surgeon, Robert Liston, from force of habit amputated the limb in half a minute flat, being accustomed to maximum speed in order to reduce the indescribable period of suffering the patient must endure. By the time Lister was performing major surgery, and anaesthesia had removed so much that was awful from surgery, post-operative infection rates were still as devastating as when he had been a student. Even in a successful year, a surgeon like Lister could still expect to see a third of his patients who had undergone major surgery die.

Over the years Lister had looked and wondered at the decomposing flesh and the suppurating sores on the wounds which his knife had inflicted. He believed, and he taught his students, that putrefaction was caused by air, or something in the air, of the operating theatre or ward, and that somehow this gave rise to wound infection. It was a confession of ignorance.

The writings of Pasteur which, in 1865, Anderson put in front of Lister were a revelation. Pasteur had already whispered in his Emperor's ear the logical line of his thoughts between putrefaction and disease. Lister now knew what the sycophantic commoner had said to the monarch. Pasteur, by this date, had come to the conclusion that putrefaction was another manifestation of fermentation in animal flesh. He believed that living organisms causing the putrefaction were carried by air. Pasteur also believed that all the micro-organisms required oxygen. Some flourished in the presence of free oxygen of the air; others acquired it from the blood, muscle, tissue or fat they were causing to turn to stinking pus.

The possibility that minute living creatures in the air might be the mischief-makers immediately made sense to Lister. It took him very little time to form the theory that, to prevent these germs gaining access to the wound could also prevent the 'hospital diseases' which followed putrefaction. He knew Pasteur had killed germs by heat, which did not help his problem. Lister reasoned that, if he covered a wound with a dressing which did not exclude air, but killed off the floating particles in it, he might have a method of reducing his death rate.

He remembered reading in a newspaper that the sewage at Carlisle had been successfully treated with carbolic acid and that the acid had, by chance, got rid of certain parasites on cattle. Lister therefore got the chemist Anderson to supply him with a sample of carbolic – German creosote, it was called: a sweet-smelling, very impure, dark liquid.

Lister chose to try out his theory by waiting for a case of a compound fracture to come into the hospital: that is, one in which the bone had pierced the skin leaving a wound in need of dressing. Cases were common, and so were complications from infection; death often followed.

The first suitable case for treatment came in March 1865. Lister tried applying carbolic acid to the wound. The technique he used, whatever its details, was a failure and presumably the wound became infected like so many other of his cases. Unusually, 1865 brought remarkably few cases of compound fractures to the Glasgow Infirmary. It was not until 12 August that the next suitable case occurred. Eleven-year-old James Greenlees was on that day carried in with a compound fracture of the left leg caused when an empty cart ran over him. Lister's treatment, highly experimental and very basic, was to smear the undiluted, impure acid onto the wound. What pain this caused James Greenlees is not recorded. Lister then covered the wound with a cloth also soaked in carbolic. He now covered the whole lot with tinfoil to prevent evaporation of the liquid. When he came to remove the dressing he found to his delight that the blood and carbolic acid had formed a firm scab with no sign whatever of putrefaction.

During the following weeks ten more cases of compound fracture came into Lister's ward. He used carbolic acid on each one. In this bunch of eleven cases there were only two tragedies. One man developed hospital gangrene in his leg when Lister was away for several weeks, and one man died from a haemorrhage when a piece of his broken bone pierced a main artery: that is, from a cause unrelated to putrefaction.

Lister was elated. Nine total recoveries out of eleven in cases such as these was an unheard-of survival rate. As usual he wrote to tell his father all about it (as ever, using the Quaker familiar form of address):

There is one of many cases at the Infirmary, *he wrote*, which I am sure will interest thee. It is one of compound fracture of the leg: with a wound of considerable size and accompanied by great bruising, and great effusion of blood into the substance of the limb causing great swelling. Though hardly expecting success, I tried the application of carbolic acid to the wound, to prevent decomposition of the blood, and so avoid the fearful mischief of suppuration throughout the limb. Well, it is now 8 days since the accident, and the patient has been going on exactly as if there were no external wound, that is as if the fracture were a simple one. His appetite, sleep, etc., good, and the limb daily diminishing in size.

The day after Lister put the first successful carbolic dressing on James Greenlees, Ignaz Semmelweis died from an infection contracted in a hospital ward. Lister was never aware of the irony. Unlike the unhappy Semmelweis, Lister knew how to publish. Lister never wrote a book in his life and putting his work on paper was, for him, a laborious procedure. It took him some time to satisfy himself that what he had done had meaning, but as soon as he had, he put it on paper and into print. His work was in the *Lancet* by early 1867. And in the same year he was addressing the British Medical Association's annual meeting in Dublin on his new antiseptic principles. There was no chance that what he had done would go unnoticed, just as there was no chance that it would not be controversial.

But Lister carried the confidence of success. He kept his father informed of his progress, letting him share in the rewards he got from the feedback. Of one woman with a compound fracture, 'the very worst case I think that I ever had,' he wrote:

today she gave me her first smile, and, being Irish, told me yesterday she had had 'an elegant night': and she made this morning her best breakfast since coming in. But this case not only throws light on the treatment of compound fracture, but on that of wounds of *all* sorts; and I have tried the new plan (or rather an improvement on the old) in the case of a gentleman from whose arm I removed two days ago a tumour, deeply seated, and such as probably would have suppurated in a somewhat serious manner with an ordinary dressing. Besides the patient is accustomed to a bottle of port every day after dinner; not a very pleasant patient to have to do with. Well his arm is today as free from pain, redness or swelling as as if it had not been touched.

Though Lister had not attempted to change the conditions of his wards, in nine months of application of his new technique he had not a single case of pyaemia, hospital gangrene or erysipelas. However, the conditions of his wards *needed* changing. Lister had discovered that, after the cholera epidemic of 1849, there had been extensive pit burials in the old cathedral churchyard adjoining the Glasgow Royal Infirmary. Coffins had been piled to within a few inches of the surface, and were separated by a four-foot-wide basement from two male accident wards, one of which was Lister's.

There were other possible sources of infection. Lister was quite clear about the principles involved in his antiseptic discovery, and he had peculiar faith in them. One German surgeon who watched him described him as the worst practitioner imaginable of the doctrine he preached. When operating he took off his coat, rolled up his shirt sleeves, pinned an unsterilised huckaback towel over his waistcoat (to protect himself, rather than the body on the table in front of him), and after using a dilute solution of carbolic to wash his hands, which had become very chapped by the repeated process, began to operate.

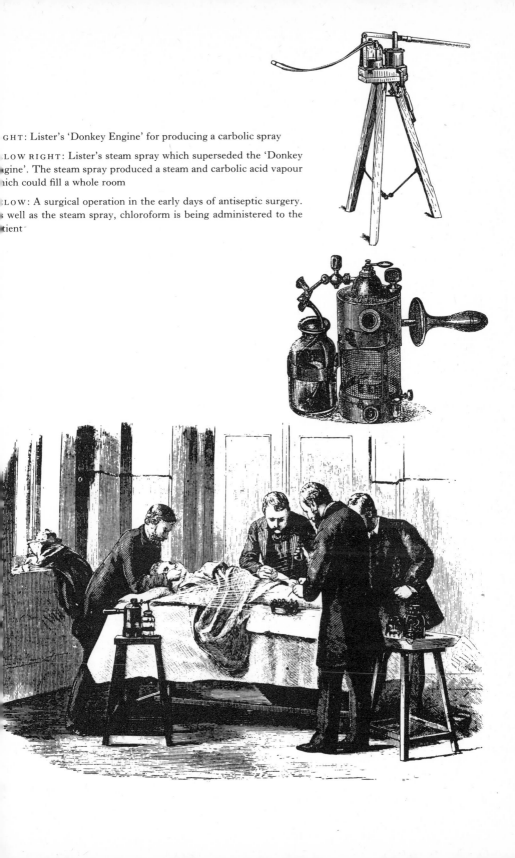

RIGHT: Lister's 'Donkey Engine' for producing a carbolic spray

BELOW RIGHT: Lister's steam spray which superseded the 'Donkey Engine'. The steam spray produced a steam and carbolic acid vapour which could fill a whole room

BELOW: A surgical operation in the early days of antiseptic surgery. As well as the steam spray, chloroform is being administered to the patient

In spite of his practical shortcomings, Lister developed his method and altered it when he saw how it could be improved, so creating the era of antiseptic surgery. He changed the form of his antiseptic dressing many times and developed antiseptic ligatures for tying up wounds and which could be left in the flesh without suppuration taking place. Soon he invented a carbolic spray which was used in operating theatres to try to kill germs in the air. It was a messy business. After a time Lister realised that the dangers of micro-organisms being carried by the fingers and instruments had worse consequences than those in the air of the operating theatre. Semmelweis, without properly understanding why, had come to the same practical conclusion twenty years earlier. But Lister had worked to his result from a position of understanding. He told his father, 'Surgery has become a different thing altogether'. So it had, though he meant *his* surgery, and with it the good he could do; he was not one for making grandiloquent claims for the profession as a whole. That was his style.

As a pathologist and surgeon Lister was workmanlike and he embraced the new bacteriology in his customary workmanlike fashion. He made useful contributions in the field after obsessively spent hours. But he had few illusions about his abilities. He wrote to his brother:

I have been labouring hard, I may say, at the subject of these wee organisms, and have made *some* progress. But though the notes fill many and many a page of the book that thee sometimes kindly wrote in, the results are small indeed.

Pasteur (whose creativity had not been diminshed by his brush with death) had few illusions about his own abilities. He rated them highly. When Lister wrote to him in 1874 to make sure that Pasteur had seen his papers in the *Lancet,* Lister ensured two things. He confirmed that his work had reached the mainstream, and he delighted Pasteur because he provided clear-cut evidence in favour of the germ theory at a time when it was still under heavy attack. After he received the papers, Pasteur replied with friendly, if superior confidence, including in his letter a few criticisms of errors in Lister's work. Gently, but firmly, as if to a sixth-form schoolboy, he wrote out a few experiments for the Englishman to try out so that he could see where he was going wrong. 'I should like to go on,' wrote Pasteur,

but writing quickly fatigues me. In the month of October, 1868, I was struck with a paralysis of the left side, and I have only half recovered from this terrible shock. My head can only stand, at one time, a limited dose of effort and work. Bending my head especially when I write is very painful to me. But I leave you with regret.

Lister had joined great company; his contribution to applied bacteriology had brought him shoulder high with the French giant. Lister's creativity was a simple process. Chance had not helped in the thing. He had read of the germ concept of disease, had formed a simple theory of antisepsis, and had applied it. The only significant piece of luck involved was the sweeping effect of the consequences. Millions of lives were saved by the new principle and what led on from it. The frightful spectre which haunted operating theatres had at last been shown to have organic substance, and Lister had demonstrated how to lay it.

But Lister was no Pasteur. His truly significant work is contained in the one principle. He had contributed an important word to the language of science. There was no page of brilliantly interconnected sentences to compare with that supplied by the half-paralysed Frenchman. On the other hand Lister did not have some of the undesirable characteristics of Pasteur. Did any of the roots of Pasteur's creativity and Lister's limitations lie in this character difference? The answer must be undoubtedly yes. Lister was not aggressive, nor ambitious, nor nationalistic, nor did he love a public quarrel. Pasteur was each one of these, and each one spurred him to push his abilities to their utmost. Lister was a gentle, kind man whose greatest reward genuinely was the relief he brought to others, and the overall good he felt he could do within his limited surgical field. Pasteur professed, as was the fashion, that his science was devoted only to humanity's ends. But in the context in which he operated the phrase has a hollow ring. Pasteur was besotted by the nature of the problems he was set and by those he set himself. They were perfectly respectable problems. His reward was in the beauty of the theory which provided the solution to the problem. And the more he could bring his aggressive nature to bear on the problem, the more channels of discovery and knowledge his imagination threw up. That the application of his work gave him pleasure and brought him reward there can be no doubt. But the more sophisticated the problems became, and the more he dabbled with the nature of life itself, the more the method of science was its own reward. Lister's simplicity, his satisfaction with the short-term, admirable, but circumscribed aim, the relief of suffering, was a measure of his creative limitations.

Lister easily lived to see his work take a central position on the medical stage, he saw it bear fruit daily – and mankind was duly grateful. None of it could have happened to a nicer chap.

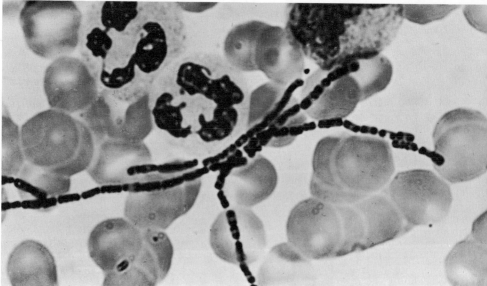

CHAPTER SIX

The germ theory now existed as a marvellous concept. Lives were being saved on the assumption that it was correct. But was it? Were the successes pure luck, and was the triumph of antiseptic surgery a simple fluke – something which worked without being properly understood? Pasteur, the chief architect of the germ theory, though he had utter confidence in it, might still see it crumble in ruins. But the matter was still not clinched. Pasteur had still not shown that any one identifiable germ caused a particular disease.

Pasteur's genius-eye-view of science covered a grand front. What was now needed was a mind of the same quality that could narrow down its new focus on this one aspect, and settle the matter. Pasteur was not yet aware that an obscure young German doctor working in a small town in East Prussia, with the right obsessive quality of mind, capable of penetrating concentration, was already at work on the problem.

Robert Koch's character was not unlike that of Pasteur. In some respects they had the same unattractive qualities of genius. Koch had all the characteristics required of a good country doctor: he was kind, took trouble, and was very patient. But he was also ambitious and quarrelsome, and took the same perverse pleasure as Pasteur in seeing himself establishing intellectual ascendency over anybody working on his own subject.

Koch and Pasteur's origins were not dissimilar. Koch too came from a poor country family with intellectual aspirants among its members. He was one of thirteen, the son of an overseer in a local mine. As a pale, quiet, myopic schoolboy it must have been obvious to his close family that the young Robert had extraordinarily mature, even peculiar, talents. Women were always to play a significant role in his life. At 15, already both intellectually competitive, and riddled with sexual guilt, he wrote:

I became, about the age of four or five, weak and sickly; I fell behind many contemporaries, as far as my body was concerned; how could that be otherwise when I squandered the noblest sap of my body, without knowing or even realising it. Only when I was twelve did I learn, quite by chance, what consequences sexual sins have.

With his tortuous worries over masturbation unresolved he fell in love with a cousin. He poured out his sexual worries, not to the girl, but to her mother, who can only have been startled by the adult introspections and confessions of the boy.

Koch, the young man, buried that passion, and found a new one at

Göttingen University: natural science. He was quite captivated by the subject. In order to learn more of it, and to indulge himself, as he had with his other passions, he had to study medicine. From the Göttingen medical school he went to Berlin and Hamburg.

Koch was 22 when in 1866 he took his first medical job in Hamburg. Following him into the city, as something of an afterthought, was a fiancée. He had surprised his family by announcing that he had become engaged to a cousin he had known since they were children together, Emmy Fraatz. The marriage appears to involve premeditation, but no passion. It has the peculiar look about it of a marriage of convenience. Plain Emmy was useful. She could look after the house, bear children, and help him in his work by acting as his laboratory assistant. It was a style to which he wanted to grow accustomed. But she was a strong character, and dominated the household she organised. Koch had dreams about emigrating to the USA; but Emmy made sure that the family's roots remained where they were: firmly planted in the country she knew.

Nevertheless, Koch's sense of duty was stronger than Emmy's demands. On the day he married her, he insisted on returning to his practice where a sick patient needed his attention.

In the first years of marriage, Koch's career did not prosper. Always in the background, attempting to push it along, was the nagging wife. The Franco-Prussian War provided Koch with a break from both his routine work as a practitioner, and from his wife. His wartime experiences influenced him in two ways – he soon became practised in the bloody procedures of the military surgeon, seeing in a few weeks a range of injuries he could never have come across in a lifetime in his small town practice: and he marched with a victorious army over French soil, and – like Pasteur – never for an instant questioned that the nation at whose disposal he was putting his science could be anything other than on the side of right.

Koch's practical experience of medicine in war put him in a position to secure the kind of stable, better-paid job that would satisfy his wife. He became District Medical Officer in Wollstein, a town in Posen with a large Polish-Jewish population. Perhaps it was as a reward that Emmy gave Koch a microscope for his 29th birthday. He could now dabble in the scientific research that was nearer his heart than medical practice. Emmy, though she did not know it, had made a decisive gesture, and one that profoundly influenced Koch's career.

He was intrigued by the new pieces of technology which were then currently being applied in biology. Already he was very adept at photography and was thinking of using it in combination with the microscope. Emmy divided the room in which he saw his patients with a brown curtain. In the back half, behind the curtain, he was able to rig up a laboratory and put in it his newly acquired hardware, his plates and slides, and the white mice and other small animals he was already beginning to use. From this place were to emerge some of the most important results in the history of medicine.

Making his living as he did in a country practice, pushing his horse on from farm to farm, learning to know what troubled the farming community, it was natural enough for Koch to begin his spare-time research into anthrax. The disease, as it had on several occasions in the previous century, was sweeping the pastures of Europe. It was a ruinous disease for the small farmers, affecting most stock, including cattle, horses and sheep, and sometimes spreading to humans. Plenty of folk-lore surrounded it. Certain areas of ground, often good soil, were said to be cursed with anthrax and to have evil spirits cast upon them. Certainly, the observation that animals repeatedly took sick on certain fields was correct. The effects were unpleasant. First there was a rise in the animal's temperature, then it would begin to tremble, gasp for breath, go into convulsions and, with the acute form, die within a few days. The name anthrax (from the Greek, coal) had been aptly coined for the disease which caused the thick flow of black blood which oozed from any scratch on the animal's skin.

At the time that Koch began to take an interest in anthrax he was well aware of the progress of the arguments for and against the germ theory of disease. At Göttingen he had been a student under Henle, whose thoughts on contagious disease had come near a basic theory. But surprisingly, as another young student, Elie Metchnikoff, discovered when he too worked at Göttingen, Henle had become supremely indifferent to any progress in the understanding of contagion. Koch's impetus came not from the German school, but from France. He knew that French scientists had shown that the blood of anthrax-infected animals contained a certain micro-organism, and that animals infected with this blood developed anthrax. This did not necessarily mean that the organisms caused anthrax, but the observation was an important starting-point for research.

Many miles away from any university centre, with no mind other than his own to direct the course of his researches, Koch set to work on the disease. The techniques he used at first were as primitive as the laboratory he worked in. To inoculate his animals he used a splinter of wood smeared with infected matter and his autopsies were carried out on the tiled surface of a stove. But the extraordinary success of the simple and, of necessity, totally original methods Koch employed is a tribute to his genius. These experiments took him three years of spare-time work. And one of the first triumphs in those three years would radically influence the techniques in all laboratories engaged in the science in which he had now involved himself. It occurred when he discovered a way of looking at micro-organisms through his microscope, at the same time protecting them from infection from the surrounding air. His method was brilliant because of its very simplicity. He made a small trough in a glass slide and put in it some of the pure aqueous humour from the eye of a dead ox, along with his infected substance. He then covered this with a second slide and sealed the edges. He could now study the life history of the tiny organisms trapped between the slides, as they fed on the pure organic fluid of the eye. He watched the organism, the bacillus of anthrax, grow into long cylindrical rods covering the whole slide. After a time, Koch was able to see spots forming: they were the thick-walled spores – like protected seeds – of anthrax.

During these three years Koch's makeshift laboratory overflowed with bull's eyes, infected mice, rabbits, slides and photographic plates. In the garden he kept a net to trap birds. But all this paraphernalia had one focus: the anthrax bacillus. He was convinced that he had isolated it. It was at the very centre of his obsessive attention: of his life. This method, refined with great care and labour over the years, was to inject mice with the germs he had grown, then pass on the infection from one to the other until, in 1876, he had worked out the complete life history of the microbe. He showed, in an experiment he was able to repeat without failure, how the anthrax spores which he got from pure cultures of the bacillus, infected inoculated animals and eventually killed them. He had demonstrated for the first time that a germ grown outside the body was directly responsible for a disease. A specific germ caused a specific disease. This, the essential postulate of the germ theory, was now proved beyond doubt in the case of this one disease.

Koch, working away in the isolated Polish border town, was as isolated a research scientist as had ever been. But he was fully aware that, if his work was to add to the useful framework of science, it must not share his obscurity; he had to publish. He sensed where he might get a sympathetic ear and some help. The director of the Botanical Institute in nearby Breslau, Ferdinand Cohn, had already published work which showed that he kept himself abreast with progress in the germ theory. Out of the blue, Koch wrote to him. Cohn had been pestered by home-chemistry-set research workers before now, and later he wrote how he mentally put Koch into this category. But he was a polite, kindly man and it was not in his nature to snub somebody, even if the intruder was most likely a rather boring country doctor riding a hobby-horse. He agreed to see Koch.

The sight of the pale, bearded young man, laden down with samples and equipment can only have confirmed Cohn's fears that he had committed himself to a couple of hours of tedium. But within a few minutes, with his first demonstrations of his techniques and results, Koch considerably impressed Cohn; so much so that Cohn looked and listened to what Koch had to say for three days. Cohn fetched in his colleagues one by one to look at what this not-so-simple village doctor could do with splinters of wood and home-made slides. They peered down his microscope and they marvelled at what they saw. They had a master in their midst, and they knew it. Within three months, Koch had written up his work, it was published in Cohn's biological journal and it was read by Pasteur. The most productive years in the history of medicine were about to begin.

With the publication of Koch's results the anthrax question ought to have been settled, but it was not. It was a disease which the French were particularly interested in since it ravaged so many of their farms. But still, in 1877, there were French biologists stating flatly that the germ present in blood and claimed as the anthrax bacillus, could be destroyed, and that the disease could be passed on without this germ being present.

At this point Pasteur moved in on the subject, and took his researches straight out to the fields: this in spite of the fact that his paralysis made it impossible for him to handle equipment without help. He set off for Chartres, where local farmers were having severe anthrax problems. At the local slaughterhouse the carcases of a horse, a sheep and a cow, all of which had succumbed to anthrax, were laid out in front of him. There he began a series of experiments designed to settle the argument.

The fact that argument was involved heightened Pasteur's interest. By now he had a solid public platform from which to argue. He had been made a member of the Académie de Médecine. There, anybody who felt the urge to pit their wits against his considerable scientific intelligence could do so. Though those who knew him well also knew that neither maturity nor partial paralysis had either blunted his razor tongue or softened his quick temper.

One of the members of the Académie who felt bold enough to face up

71

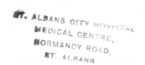

to Pasteur in the middle of his anthrax researches was the Professor of the Alfort School of Anatomy. A ponderous, but well-meaning individual, Professor Colin had, over the years, stuck his neck out in favour of spontaneous generation, and now made it known that he doubted the validity of the new work on anthrax. He also let it be known that he had carried out more than five hundred experiments on anthrax in the past twelve years.

At this news Pasteur bristled. He himself had by now cultured the anthrax bacillus in a flask of urine, sown a drop from the liquid into a second vessel, sown a drop from that into a third, and so on for a hundred seedings. With his last seeding he had still succeeded in giving animals anthrax. Yet Colin persisted in asking some pertinent questions. Why did the disease behave as it did? Why was it only found in certain areas? Why only in the summer?

In July 1877 Pasteur, in his sweeping, even smug, manner, announced that it was not possible to give anthrax to birds. Colin flatly contradicted him. For Pasteur this was the unmistakable shout to join battle. He manoeuvred his forces with care. He prepared, parcelled up, and sent Colin a culture of anthrax bacillus and, in exchange, asked that Colin should use it to send him a hen suffering from the disease.

March 1878 arrived and Pasteur had had no reply. Typically, he chose to describe to a full meeting of the Académie de Médecine how he had routed Colin. Pasteur spared Colin, who was present at the meeting, nothing. 'I saw Monsieur Colin coming into my laboratory,' Pasteur told his audience of Academicians. 'And even before I shook hands with him, I said to him: "Why have you not brought that diseased hen?" – "Trust me," answered Colin, "you shall have it next week."' Pasteur was playing the situation for laughs. The wretched Colin had, of course, never turned up with his hen. When he stood up and began to explain that he had tried the experiment, but had regrettably failed to close the hens' cage properly, and had had them mauled by a dog, he was a lost man. The Academy was in fits of laughter.

But Pasteur had by no means finished with Colin. He did a sharp about-face. He told his audience how he had gone on to say to Colin, 'Well, my dear colleagues, I will show you that it *is* possible to give anthrax to hens'. He had every intention of demonstrating to the Academy Colin's complete incompetence. A few days later he left his laboratory carrying a cage with three hens in it, one of them dead. He got into a cab, ordered it to the Académie, and deposited the cage on a desk. The prostrate hen, he announced, was dead from anthrax.

Blasé about the superiority of his own powers of reasoning, Pasteur had seen how to meet his own challenge, and realised that there was little risk that Colin would ever be capable of the same piece of intuitive science. Pasteur had noticed that the temperature of a hen's body is higher than that of most animals. He had therefore taken an inoculated hen and cooled it by partly immersing it in water. The white dead body of the hen now lay on the desk, with the unmistakable signs of death from anthrax, contracted when Pasteur had lowered its temperature. In a brilliant series of experiments he had discovered that a fall in temperature of a hen's body from $42°$ to that of a rabbit at $38°$ was enough to make it susceptible to anthrax.

The scene at the Académie demonstrated two things. First, Pasteur's utter mastery of scientific technique which he could call on to support his extraordinary scientific intuition. Second, the driving necessity he felt to demonstrate his abilities publicly, irrespective of the humiliation he left in his wake.

As a result of this brilliant piece of self-publicity the Ministry of Agriculture asked Pasteur to look into the anthrax problem in the Eure-et-Loir district. Pasteur was becoming well known outside scientific circles. The local farmers now welcomed the little man who dragged a foot and dangled a loose arm. A great deal was expected of him. He did not disappoint. He fed alfalfa infected by anthrax spores to animals. When this had little effect on the mortality he mixed thistles with the fodder in order to puncture the beasts' mouths and tongues. Mortality suddenly increased. Thus, with this very simple experiment, Pasteur showed how the disease was spread in the first place: through open wounds in the mouth.

But the question remained, why in summer? And why in dry seasons? Why often in one place and not in others? Pasteur now looked for and found anthrax spores in the soil in these infected regions. Then, on a farm near Chartres, he had a sheep buried that had died of anthrax. Ten months later, he found that the soil was contaminated. Soon he had enough evidence to show that the spores were carried to the surface by earthworms.

Eventually he was able to feed earth contaminated with anthrax spores to earthworms, open their bodies a few days later and withdraw from their intestines tiny cylinders of soil containing anthrax spores. When the worms deposited their turnings in dry weather, the dust containing spores was blown on to plants, then to animals' mouths. It was a wonderfully pastoral solution to a pastoral problem and Pasteur was again able to impress the farmers with clear instructions for preventive measures: avoid foodstuffs which might cause skin damage and, most important, burn not bury anthrax-infected animals. To these practical men, Pasteur was the practical scientist *par excellence*.

While Pasteur was holding centre stage in a drama of his own making at the Académie – the institution at the intellectual centre of French science – Robert Koch was still shuffling his mice, bulls' eyes and slides on the crowded work-bench behind the brown curtain in his small-town consulting room.

Koch knew all about Pasteur's doings; they were news by now, and he was worried. Pasteur was claiming originality for some experimental results and observations which Koch himself had discovered and published. He began to watch the Frenchman with a wary eye, and was now determined to find recognition for his work in the highest places. He took it to one of the most famous of contemporary European figures: a man then held in almost God-like esteem in medical circles – Rudolf Virchow.

Virchow worked at the hospital whose reputation he had helped create: the Berlin Charité. He was a notable politician, having been leader of the Opposition in the Reichstag, an archaeologist, having dug with Schliemann at Troy, but above all a pathologist, distinguished for his studies of cell theory and disease. A blessing from Virchow could have meant all the difference to Koch's career, and Koch knew it. But his visit to the great man was a disaster. It was an unbearably hot day. Virchow was in a bad temper, and his judgement was crushing. He turned over a few pages of Koch's work on anthrax and muttered, 'The whole thing seems highly improbable'. Thus brutalised by one of Germany's greatest living scientific intellects, Koch spent that evening in an atmosphere far removed from that of a laboratory or university – in a music hall. His relaxations, like his women friends, were unintellectual. But he was far from destroyed by his experience. He had other work under way beyond what he had shown Virchow.

He was developing methods of staining micro-organisms with dyes and novel ways of looking at them by applying microscope and camera technology. Using new Zeiss lenses he was able to see and photograph tiny bacteria whose presence had never before been recognised. His refined abilities, concentrated on one particular problem, here made it possible for him to make yet another major contribution to the understanding of the nature of disease. He infected mice with putrefying blood. Nothing was easier than to make the mice succumb to an experimental disease: a septicaemia. Whatever caused the disease was clearly extremely virulent. But when Koch first looked down his microscope at septicaemic blood he could see no bacteria whatever. He was convinced however, that the agent responsible for the disease was there in the blood, and that it should be possible to make it visible. At last he hit on the idea of using an Abbé condenser with his microscope. This combination of lenses and illumination of the object being studied now made it possible for him to recognise the micro-organisms in the blood with absolute certainty. He dried it and stained the bacilli present with methyl violet. Although the organisms were minute – about 1 micrometer (micron) long and 0·1 micrometer thick – he was able to see large numbers of them among the blood corpuscles.

To crown his technique, Koch repeated his experiment several times, passing infection from an original fluid containing a mixture of bacteria from one animal to another. With enormous experimental skill he was able to separate the bacteria during the chain of passage, so that by the time he reached the last animal in each chain only one type of micro-organism remained and there could be no doubt that the disease from which the animal was suffering was caused by the organism. Thus, using animals to passage infection artificially, Koch now had an exceptional method of pure cultivation of bacteria. Using it, his success was truly a monument to his skills as a scientific practitioner. He identified the micro-organisms responsible for common types of septicaemia, gangrene, abscesses, pyaemia and erysipelas. These were some of the diseases that had blighted the lives of victims of surgery before Joseph Lister began to apply germ theory to his hazardous profession. Now Koch had produced the detailed information which showed why Lister's antisepsis was successful.

Koch's scientific method was by now so predictably successful that recognition in some form or other had to come. The friends he had made at the University in Breslau tried to engineer him into a professorship, but their efforts failed. When, in 1879, the post of Town Medical Officer in Breslau came vacant, Koch jumped at the opportunity of leaving Wollstein and anonymity. He would at last have the stimulation of University minds, even though he would not be part of the University. Emma and their daughter would have a new house, and they would be released from the primitive country existence they had led for the past half dozen years.

The job and the dream lasted two months. Koch did not make out in Breslau. There was no private practice and no money to support the family. Koch crept back to Wollstein, to the country folk who respected him, and to the work behind the brown curtain. Koch still had not found a scientific audience to which he could play.

Pasteur had *his* audience. It was captive; something in his personality held it. He had a cause, and he had now found a catch-phrase: 'Seek the microbe'. The word had been invented by an ardent admirer of Pasteur, Dr Sedillot, the aged, now retired director of the French Army Medical School. Pasteur liked the sound of 'microbe', and now he used it on the banner he was bearing. 'Seek the microbe', he told all the young workers in his laboratory. And they did.

Pasteur was beginning to spend as much time in hospitals as in his laboratory. This was in spite of the fact that he was vastly over-sensitive to sights that could be seen every day in hospital beds. A glimpse of a lanced boil made him wince, and watching autopsies, which he frequently did, sometimes made him ill. Nevertheless, he relentlessly followed what he clearly saw as the logical path. It led him to study some of the most infamous of killer diseases of his time.

Puerperal fever was still vivid in the memory of many Parisians, even though it was now twenty-five years since it had swept the Paris Maternité Hospital. The young Dr Tarnier, who had registered the 31 deaths out of 32 confinements in the first ten days of May 1856, had never heard of Ignaz Semmelweis's work in Vienna. When Tarnier subsequently heard of Joseph Lister, he got hold of carbolic acid and began to employ it with sweeping success in the Maternité.

During this period Pasteur began to take an interest in the disease. His work was as fruitful as ever. Although his findings were still premature, when the opportunity came to reveal what he had discovered he grabbed greedily at the chance to stage another drama. It came one day at a meeting of the Académie when one member was sounding-off unformed ideas on the causes of epidemic diseases in maternity hospitals. Pasteur sensed the basic ignorance of the speaker. He was unable to resist the temptation to expose the argument's flaws, and to let the assembly glimpse the flash of his own brilliance. From his seat he interrupted. 'None of those things cause the epidemic. It is the nursing and medical staff who carry the microbe from an infected woman to a healthy one,' he shouted. It was the same sentence Ignaz Semmelweis had shouted until his death, and no one had listened. When Pasteur shouted, all listened. But he could go far beyond Semmelweis. He limped down to the blackboard and drew something on it that looked like a string of beads. It was what is now known as the *streptococcus*. 'There,' he said, 'that is what it's like'.

If the word luck has any meaning in the personal and scientific affairs of Louis Pasteur, then he chased and pushed that luck wherever he wanted it to go. Koch had not yet acquired the knack. However, when fortune did appear, offering to take him out of his pastoral backwater, he welcomed it. In 1880 he was offered a job in the Imperial Health Office. He accepted with almost indecent haste. Within three days of the arrival of the cable confirming his appointment, Koch had packed and left for Berlin, dragging Emmy after him and leaving half their belongings behind.

As when the young Pasteur arrived in Paris, the laboratory facilities provided for Koch were primitive in the extreme; a small room with a small window was his whole empire. But it was all he needed; it was all he had ever been accustomed to. What took him quite by surprise in its bounty was the provision of two young assistants, Friedrich Löffler and Georg Gaffky. At first Koch had not the slightest idea how to use them or suggest how they might profitably spend their time. Soon, however, he was astounding them with the bacteriological techniques which he, all alone, had developed in the back of his Wollstein consulting room. These two young men were the first of a whole school of research workers on whom Koch would have profound influence. He showed them how to boil, slice and prepare potato on which to grow pure cultures of bacteria. He had found this to be a wonderful solid substance on which to study a germ, far superior to the uncertain method of trying to isolate it from a swimming, quivering mixture in a drop of liquid. This devastatingly simple technique would revolutionise the experimental practice of biology. Some believe Koch's coronation of the potato to be the most important single technical discovery in the whole of the history of the science. But important as even this was, Koch used it as a stepping-stone to yet another technique. Soon he perfected a method using agar-agar jelly as a solid in which to suspend his bacteria. Again his source of inspiration was mundane; as with the potato, it was taken from Emmy's kitchen. The method was suggested by the wife of one of his colleagues whose mother had been using agar-agar to make her jellies.

With the move to Berlin the oyster of the world was just beginning to open to Robert Koch; the germ theory which he could operate on so deftly was its pearl. His work was now becoming well known, and in 1881, when he was invited to speak at the Seventh International Medical Congress in London, in terms of scientific recognition, he had arrived. There, with the help of a magic lantern and his photographic slides, he was able to show off his methods of getting pure cultures of microbes from solid media.

Louis Pasteur was present at the lecture; so was Joseph Lister. The room at King's College that day contained the three men who, in their various ways, had virtually created, established and utilised the germ theory. They had put into motion a civilising force that would have its effect on every succeeding generation. But civilising though the force was, it did nothing whatever to unify either science or nations. It was only ten years after the Franco-Prussian War and enmities still ran deep. At dinner at his home, Lister was forced to invite French and German delegates on different days. Lister knew of Pasteur's bitter and whole-sale condemnation of the German nation. He also knew that Koch had served with the victorious armies. Nevertheless, gentle friend-of-all Lister did succeed in getting the two to meet. The occasion ought to have been one of the great moments of amity in the history of science, with all the romance of a Livingstone-Stanley confrontation. Instead it was all a ghastly anticlimax. Pasteur took Koch's hand and muttered, 'That's great progress, Monsieur'. These few words represented the height of the relationship between two of the most exceptional men of science of their age. From then on it went only pathetically downhill.

But no matter what their personal differences, these two founders of modern bacteriology were agreed on one thing: that there existed microbes with different characteristics and that each was responsible for a specific disease. But not all scientists, and by no means the whole of the general public, were convinced of the function, or even of the existence of these creatures. 'There!' wrote one sceptical scientist in a newspaper, after one of Pasteur's publications, 'One more microbe; when there are a hundred we shall make a cross'.

Pasteur was not in the least moved by the emotional arguments that preferred establishment medicine to the new science. Already he was involved in the hunt for more of the creatures, and was slowly working himself towards familiarity with more human diseases. But there was still much to be learnt from animals. In the farming communities he now knew so well, Pasteur had come across another disease, and again his sensibilities were revolted by the suffering it caused. The chicken cholera that was destroying great flocks of domestic birds had a cruel effect on fowl. It first made the bird's feathers stick out from its body, forcing it into the shape of a ball. It would stagger along until it was cast into the deep sleep which preceded death. It was a sad sight, but the pragmatic farmers of the poultry farming regions were more saddened by the effect on their annual turnover than the suffering of their birds. As ever, concern over profits moved mountains of interest in the new science of bacteriology.

Pasteur was sent the head of a cock that had died from the disease, and was soon able to find a way of culturing the microbe in a chicken broth. If he put a few drops of his culture on to bread, or meat, and fed it to chickens, the disease spread rapidly from the intestines of the birds, to their droppings, then on to the rest of the flock.

When he began to work on chicken cholera, Pasteur had already gathered to his laboratory a number of highly talented assistants. Charles Chamberland and Émile Roux were two of the typically brilliant young men who were delighted to spend their days in working conditions which, some of them said, were those to be expected in a monastic order rather than in a biology laboratory. A single-minded devotion to the hunt for the microbe was expected.

That year, 1880, was warm and dry. It was high summer when an incident of neglect intruded on the monastic routine. Pasteur had left instructions for Chamberland to inoculate a batch of hens with some cultures of the chicken cholera bacillus. For some reason, whether or not he was in a rush to get off on his holidays, Chamberland did not carry out the inoculations before the laboratory closed down for the long vacation. When Chamberland returned from his holiday he continued the experiment where he had left it. The experimental results, however, did not follow the earlier pattern. Chamberland had to report to Pasteur that though the germs at first made the chickens ill, they quickly recovered and remained hale and hearty in spite of his best efforts.

The obvious assumption to make now was that some error had been made in the preparation of the culture. Chamberland was in the act of throwing it away when Pasteur stopped him. This decision on the part of Pasteur was one of unanalysable genius: an intuition plucked from the recesses of his mind, and in which one can see logic only in retrospect. Pasteur told his assistants to make a new batch of virulent culture and inject it into the same birds. They did as they were told. The result was that the chickens not only survived, they survived healthily. A new batch of chickens straight from the local market and injected with the new culture succumbed, as all previously had done, to chicken cholera.

Chamberland had stumbled across one of the greatest discoveries of his life, and if Pasteur had not been watching him, Chamberland would have had no idea what it was he was tripping over. Pasteur's guess, based on his observations of animals which survived epidemics out in the fields, was that the microbes left exposed during the summer vacation had become attenuated: weakened by their contact with the oxygen of the air. When injected into hens, these microbes somehow gave protection from chicken cholera.

A bacteriological link had been discovered with the work of Lady Mary Wortley Montagu and Edward Jenner. Pasteur, like Jenner, had discovered a vaccine. In honour of Jenner he used the word vaccination for the process. Even though like Jenner, Pasteur had no idea why the vaccine gave immunity to chicken cholera, he had taken the first step towards putting immunology on a scientific footing. The ultimate economic effects of this work, which came about through the worries of a few chicken farmers over a few francs, would in world terms be enormous. This piece of intuitive, creative science is all the more breathtaking when it is seen how Pasteur pin-pointed the implications of an apparently irregular experimental result while his co-workers whose errors had produced these results would have shrugged and moved on.

But he was a peculiar man to admire. His bursts of emotion were now, in sick middle-age, becoming more uncontrollable. Friends tried to persuade him to be less vitriolic in his cruel public demolition of any scientist who tried to oppose his views. For a time he listened and controlled his tongue. But in moments of stress he was quite incapable of keeping a grip on himself. It touched absurdity and bathos when one day at the Académie de Médecine he provoked an eighty-year-old member into an argument over vaccination in such a way that the old man, Jules Guérin, keeping up with the ludicrous fashion of the time, challenged Pasteur to a duel. The thought of himself, an ageing paralytic, facing the sword of a gentle geriatric doctor at dawn in the Bois de Boulogne, was too much even for Pasteur. Meekly, he withdrew. The experience, however, left him no better equipped to control his emotions in public.

The chicken cholera experiment was another plank for Pasteur to add to his framework of scientific logic. He was now determined to apply the same principles he had discovered in chicken cholera to anthrax. But the anthrax bacillus provided him and his assistants with a serious problem. Far from weakening the microbe, exposure to air simply allowed anthrax spores to go on forming – and the spores were responsible for the disease's perniciousness.

Therefore, what Pasteur had to do was to prevent spore formation. His assistants found that anthrax bacteria stop multiplying at 45°C. But at 42–43°C spores are no longer produced. So, by keeping cultures at temperatures of between 42° and 43° for long periods, Pasteur reasoned it should be possible to keep bacilli alive but at the same time prevent spore formation. Maintaining these conditions for eight days Pasteur was able to produce live anthrax bacilli which had lost their virulence. He now found he could successfully vaccinate guinea-pigs,

rabbits and sheep with his weakened microbes.

In January 1881, a month before the discovery of the anthrax vaccine, one of the editors of *The Veterinary Press,* a Monsieur Rossignol, had written a skit on the fashionable germ theory: microbiolatry he called it; its Pontiff, said Rossignol, was Pasteur whose sacramental words were, 'I have spoken'. Rossignol may not have understood the formidable nature of the methods he was taking on, but he had no doubts about the man. When he suggested that a subscription be raised for funds to enable the anthrax vaccination to be put to a searching practical test, Pasteur could not have been more receptive. The proposal meant that large numbers of animals and facilities would be made freely available to him. And the idea of a practical public challenge to his abilities was guaranteed to elicit an energetic and pugnacious response.

The experiment was a large and well-publicised affair. It was to be more grand than any other of the scientific battles Pasteur had been involved in, and was therefore all the more acceptable to him. It took place at Pouilly le Fort, near Melun, on a farm owned by Rossignol. Leading Paris newspapers sent their observers, and so did *The Times,* in the impressive shape of Mr Henri de Blowitz, its larger than life Paris correspondent, once dubbed by *Punch* as Blowitz-own-Tromp. De Blowitz would soon make Pasteur world news.

The conditions of the challenge to Pasteur were spelled out to him by the benign old Chairman of the Melun Agricultural Society, Baron de la Rochette, but they had been designed by Rossignol, who hoped to see the theory of vaccination speared by a thorough practical test in front of plenty of witnesses. Sixty sheep were presented by the Society for Pasteur's work. Twenty-five were to be inoculated with anti-anthrax vaccine from Pasteur's laboratory, two doses to be given at 12- or 15-day intervals. A few days later twenty-five more sheep were to receive, along with the vaccinated batch, a virulent anthrax culture. The survivors, if there were any, could then be compared with the remaining ten sheep. La Rochette asked for a few cows to be included in the experiment. Pasteur, showman that he was, agreed to take on ten cows, although, as he pointed out, his work on cattle had not reached the same stage as that with sheep.

The date fixed for what was to become one of the best-known experiments of all time was 5 May 1882. Word had got around. It was as theatrical a setting as Pasteur used for his arguments in the Academy. Here he had the open stage of the countryside, and the possibility of his biggest audience. From the stations at Melun and Cesson, a whole stream of farmers, laboratory workers, vets and doctors moved towards Rossignol's fields at Pouilly le Fort. Limping along with them was Pasteur, preparing to direct operations, but too crippled to be able to handle the injections himself.

The animals were separated into the agreed groups in a large shed. Two goats had been substituted for a couple of sheep, and an ox for one of the cows. The mixture of species, humans as varied as the animals, created an atmosphere more of circus than of sober scientific experiment; Pasteur was the ringmaster, whipping-in his assistants, bearing their needles, to the animals' thighs. But over all there hung an atmosphere of intense seriousness.

Chamberland and Roux knew that this experiment was trickier than even the scientifically informed members of the audience guessed. The techniques involved were formidably difficult to carry out successfully in the laboratory, let alone in a field. The vaccine they were using was one prepared from microbes which had been attenuated, not by the tricky process of heating at 42–43 °C, but by treating with potassium bichromate. It was a new method they had developed, and were now applying away from the controlled conditions of the laboratory for the first time. They dared not work with their original vaccine, so susceptible to experimental errors, in the risky conditions in the open.

Twelve days later the crowd of onlookers and the menagerie of animals gathered again under the same shed, and the first batch of animals were given their second dose of vaccine. On Tuesday, 31 May, the whole bunch of animals, vaccinated and unvaccinated, were ready to be treated with a virulent strain of anthrax which Pasteur had grown in his laboratory. Just before the inoculation was to take place, a local vet pushed his way through the crowd, grabbed the tube containing the liquid and gave it a vigorous shake to make sure it was thoroughly mixed. He had been put up to the act by Pasteur's enemy of old, Colin, who had said that the liquid was in two layers: the top layer inert, and the bottom layer virulent. Colin hinted that, by drawing off fluid from the top layer, Pasteur could make sure that the vaccinated sheep would not contract anthrax.

Pasteur watched the antics with the superiority of a man who was too clever a scientist to cheat in this fashion. Had he needed to fake his results he could have devised a less chancy method. But, to improve the drama, and knowing it would make no difference whatever to what he was doing, Pasteur agreed to give triple doses of virulent anthrax injection to the animals, and to inject vaccinated and unvaccinated animals alternately.

Farmers and scientists now dispersed to their respective occupations to await the morbid results: the death of half or the whole of the animals, and the life or death of Pasteur's reputation.

A few days later Chamberland and Roux took the train back to Pouilly le Fort and began to examine the sheep. There was no doubt about the virulence of the anthrax strain they had used. Several unvaccinated animals were ill. So too, however, were a number of the vaccinated sheep; they were standing in the field with drooping heads, were breathless and, as the two assistants found from their thermometer readings, had raised temperatures. They went back to Pasteur and, in an uneasy interview, watched him react to the news. The worst side of his many-faced temperament surfaced. The reaction was unpleasant. Pasteur turned on Roux and threw the blame directly on to him, telling him to take himself back to Pouilly to face the inevitable humiliation. Pasteur spent that night anxiously worrying himself into a nervous state: it was a side of his personality he kept hidden on those public occasions when he demonstrated his superior powers.

Next morning, however, a telegram arrived which restored his equilibrium. It said simply that the vaccinated sheep were well. On 2 June, the date fixed for the assessment of the experiment, at the agreed time of two o'clock, Pasteur walked into Rossignol's farm. Behind him were the young men whose skill he had nurtured, then doubted. As he appeared in the farm-yard, a subdued ripple of applause began which soon burst into a cheer. Pasteur knew he was a success. The carcasses of 22 unvaccinated sheep were laid out in a row; two more were standing by on the point of death. To oblige in the drama, they died as the spectators watched. Every one of the vaccinated group were grazing, unconcerned by either the deaths of their kind, or the excitement of the gathered crowd.

But the numbers who heard the shouting were much bigger than the gathering round the shed at Pouilly le Fort. Pasteur had gambled his reputation in a way that ensured that the whole of Europe knew what he had risked and achieved. The following week he stood in front of the Académie des Sciences to list the full details of his success, and to make sure that it went accurately into print.

The one man who, before all others, should have seen that what this despotic, superior little Frenchman had produced was the promise of an immeasurable advance in the condition of mankind was Robert Koch, a man still in his thirties, whose abilities were on the rise. But Pasteur, nearly sixty, who, time and again, had shown his preeminence in the scientific profession, had now trampled on the young Koch's own ground. Anthrax was Koch's subject; the work he had done on it had made him.

It so happened that the territorial imperatives of both men were unusually highly developed. Koch's dismissal of what Pasteur had done was not simply haughty, it was thoughtless. It even bordered on mockery of one who, after all, was an impeccably distinguished senior member of the profession. In effect, he said that Pasteur had contributed nothing new to science, and had not really done what he had claimed.

By now Koch had other things on his mind besides anthrax. Like Pasteur, he wanted publicity for his new ideas but, since he believed that Pasteur had stolen so much of his fundamental work on anthrax, he wanted his sole ownership of his new concepts to be unmistakable. For, what he was now concerned with was of first-rank public importance and scientific interest. He therefore set about his experiments in strict secrecy, not even telling some of his closest friends what he was up to. This time, when he revealed all, it would be in his own good time.

He had decided that there was a good chance that he could determine the cause of a human disease: tuberculosis. This, of all diseases, needed to be tackled. It not only caused the deaths of millions, infecting whole communities, it had associated with it an important touch of romance and even glamour. Tuberculosis brought neither the ghastly symptoms, nor necessarily left behind any of the awful disfigurement of other killer diseases. The fact that Keats, Chopin and la Dame au Camélias could make stately, even beautiful, deaths in the clasp of 'the white plague' made it an interesting, even acceptable condition: so much so that the pale, gentle and weak disposition of the consumptive was fashionable. Yet the totality of tuberculosis was in no way romantic. For many years it was the chief cause of death in the United States, taking a particularly wicked toll on the black population. Its effects on families (the Brontës are a romantic example) had once suggested that there was a genetic cause to the disease. Damp housing had been popularly thought to be responsible.

When Koch returned to Germany from the 1881 International Medical Congress in London the infective nature of tuberculosis had been understood for fifteen years. A Frenchman, Jean-Antoine Villemin, had shown how it could be inoculated from man to rabbit. Koch intended that the next step towards an understanding of the disease – and the next progressive step in the germ theory – should come from his laboratory in Berlin. With his assistants now using the advanced techniques he had taught them, and with Koch himself inventing new

85

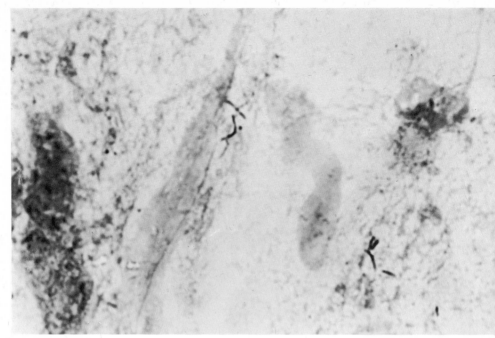

ABOVE: Bacilli of tuberculosis in sputum magnified approximately 1800 times

RIGHT : Test-tubes of tuberculosis bacilli culture as prepared by Koch

methods to cope with the challenge he had set himself, the work sped along in secret. Considering the newness of the subject, Koch's progress was miraculous. There was none other to compare with it, except the similar bounding intellectual stride of Louis Pasteur.

Koch's main problem was that the tubercle bacillus is extremely small, about a third the size of the anthrax microbe. The fact that the bacilli congregate in small numbers, and are difficult to see, meant that in the past workers had missed seeing them altogether. He had to devise ways of staining the tiny microbe so that he could make it visible against the tissues surrounding it. He had to prepare sterilised sheep's blood serum to grow the bacillus. And since the tubercle bacillus takes weeks to grow, he had to wait with infinite patience until long past the point where other workers would have thrown the stuff away and started again. Finally, with the pure culture he had succeeded in growing, he had to inject healthy animals and show that the organism that killed them was the one he had separated. On 18 August 1881 Koch inoculated his first guinea-pig with tissue from an ape that had died from the disease.

By May 1882 Koch was ready to present his findings to the Berlin Physiological Society in one of its small reading rooms. The place held only seventy-two chairs and a table on which Koch had spread out 200 of his preparations. One of the distinguished audience which crowded in past Koch's obsessively ordered spread, as he pushed through remarked that it looked rather 'like a cold buffet'. But Koch was never better at impressing others than when he had a series of experimental results to show off. Though his style of lecturing was halting and poorly delivered, what he had to say was rational, clear and tremendously impressive to those present who understood its implications. For the first time the cause of an infectious disease in man was certain.

After he finished speaking there was a long silence. Then a hubbub of discussion broke out. One of the scientists there, an untidy young man who chain-smoked cigars, quickly dashed from the room to start to try the experimental techniques he had seen for himself. He was Paul Ehrlich, a pupil of Koch. That disappearance from the room that day had its counterpart in physics sixty years later. In Washington, a few months before the opening shots of the Second World War, Niels Bohr lectured the American Physical Society on the mechanism of nuclear fission. Before he finished speaking, several physicists left the room to verify what he was saying in their own laboratories. The fact that both these experimental results were the first critical steps in man's ability to control the lives and deaths of large numbers of his fellow men had been appreciated by the discerning few on both occasions. Both results would have profound international consequences. In both cases the application of the science involved would determine the nature of political and economic decisions on a grand scale.

In 1882 only optimism would rise from an understanding of the combined work of the bacteriologists, the germ seekers, the microbe hunters. It should have been a year of concord and celebration within the house of science. The amity might have been publicly displayed when Pasteur and Koch again met face to face at the International Congress of Hygiene in Geneva. Instead a freezing hostility gripped the gathering, which had its origins in the antagonism between these two men: Koch, stony faced and secretive, guarding his pieces of scientific creativity with a childish possessiveness, reluctant to distribute credit anywhere beyond his own doorstep; Pasteur, with cruelly twisted features, imperious, sarcastic, overtly anti-German, and over-sensitive to any criticism of his scientific abilities.

It was at this conference that Koch left Pasteur in no doubt that he, Koch, believed Pasteur had contributed nothing new to science. Pasteur's uncontrollable emotions again flowed over and he rushed into his battle position. Immediately he challenged Koch: demanded a debate, a face-to-face confrontation, to bring to a head the years of dislike at a distance. This aggressive manner was the only way in which he could conduct his scientific business once his sensitive skin had been pierced. But Koch would have none of it. Pasteur had to sublimate what was now his hatred in print. He spent Christmas Day of 1882 writing a letter to Koch which had in it no drop either of Yuletide spirit or scientific brotherly balm. Indeed, Pasteur had probably waited until 25 December, the more dramatically to headline his invective. In his letter, Pasteur was at his most biting and launched into a full-scale verbal war in an attempt to establish his own priority and superiority.

This period of four or five years from 1877 had been one of the most creative in the history of science. The germ theory was now established too firmly to be seriously shaken down. The science of bacteriology seemed entirely humanitarian in its applications since all was now set fair for conquering human disease. But to look for fine human motives in this glorious work is, surprisingly, to look for something conspicuous by its absence: something created after the event. Pasteur and Koch were both entranced by the problems they set themselves. To recognise the problems at all was an achievement in itself. Pasteur's was a great conceptual scheme. Koch's wonderful facility with method and technique spread its base securely and made applications possible.

88

RIGHT: The 'comma'-shaped bacilli of cholera magnified approximately 2000 times

BELOW: An advertisement which appeared at the time that an epidemic of cholera was raging in Hamburg in *The Illustrated London News*, 1892, claiming that cigarettes could prevent, perhaps even cure the disease

OVERLEAF ABOVE: Disinfecting luggage during a cholera epidemic in the south of France, 1884

OVERLEAF BELOW: A women's hospital set up during the same epidemic

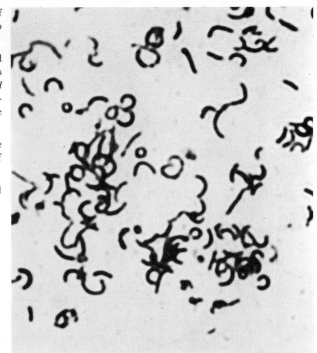

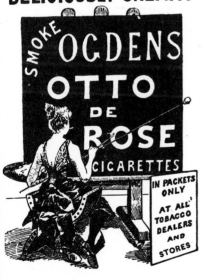

DELICIOUSLY CREAMY

SMOKE OGDENS OTTO DE ROSE CIGARETTES

IN PACKETS ONLY AT ALL TOBACCO DEALERS AND STORES

CHOLERA.

IMPORTANT DISCOVERY

BY
DR. TASSINARI.

Dr. Tassinari has carried out a series of experiments with tobacco smoke upon the germs of various infectious disorders, with results which are satisfactory.

Smoke passed through a hollow ball containing Cholera bacilli had a fatal effect upon germ life.

OGDEN'S
OTTO DE ROSE
CIGARETTES,
PURE VIRGINIA,
FIRST QUALITY,
PURE RICE PAPER,
ARE UNRIVALLED.

But solving the problem was an end in itself and it was the main end. Finer motives were very secondary, and not important spurs to problem solving. There is no real evidence that either the love of man or the scientific brotherhood were inspirational urges driving on the two characters in the van of the scientific advance. There were romantic pictures invented both by contemporaries and by later generations who saw the humanitarian end as the product of high motives. Between Koch and Pasteur there was rank dislike. What speeded on the process of discovery and creativity more than anything else was blunt competitiveness, harsh national rivalry and a single-minded passion for priority.

But if the motives seem unpleasant, the achievement was staggering. The advances so far now showed how much more could be uncovered. The bacteriological gold-rush was on, and the two main teams, having staked their claims, were waiting to dig.

No attempt was made to hide the fact that a race was in progress. In 1883 the situation had developed to one of direct conflict. A severe cholera epidemic had broken out in Damietta, in Egypt. Both Germans and French had been waiting for an opportunity to study the disease and when Koch and his team arrived in Alexandria in August, he found that a team sent by Pasteur, and led by Emile Roux, was already installed in a hotel in the city.

The Germans quickly organised themselves and launched an attack on the problem, getting hold of cholera victims, carrying out autopsies and examining the intestines and stools of victims. Koch already suspected that a bacillus shaped like a comma was the cause of the disease. It now came as a great blow to his pride in his own high-speed technical abilities when he heard that the French were packing their bags. He was told that the Frenchmen had done what they had set out to do: identified the causative agent of cholera.

It was, therefore, a humbled Koch who had to go to Roux and Thuillier, another of Pasteur's young assistants, and ask for permission to look at what they had found. Koch's humility was short-lived. He soon saw that the Frenchmen had mistaken platelets – particles which are normal constituents of the blood – for the cholera bacillus.

But Koch could not revel for long in this *Schadenfreude*. For about a fornight the epidemic waned and there was scarcely a case of cholera with which the teams could occupy themselves. The French team relaxed and kept themselves cool with swims in the sea. At three o'clock in the morning of 15 September Koch was wakened with the news that Thuillier had been taken seriously ill. He had been infected by the water-borne bacillus he had failed to recognise in his work. He was dying.

Koch did as Roux asked him and went to Thuillier's bedside. The young Frenchman looked up at the German and asked, 'Have we found the cholera bacillus?' Koch, to his eternal credit, lied. 'Yes,' he replied. Thuillier died shortly afterwards.

Roux wrote to Pasteur telling him how well Koch and the German team had behaved: how they had nailed simple laurel wreaths on Thuillier's coffin, and spoken fine words of praise. Pasteur wrote, 'I can only console myself for this death by thinking of our beloved country and all he has done for it.'

CHAPTER SEVEN

Pasteur had vivid powers of recall, even as an ageing, sick man. It was one of the characteristics that made him such a successful scientist. One of the scenes from childhood which he remembered with great clarity had taken place in the smithy near the house where he was born. There he had watched a man who had been bitten by a rabid wolf having his flesh cauterised with a red-hot iron where the wolf's teeth had pierced and where saliva had entered his limbs. The young Pasteur had witnessed a terrible cure, but one no more terrible than the effects of the disease: the period of waiting for the first sign, the violent behaviour, the contraction of the muscles of the throat, the dread of the sight of water, and the convulsions leading to death. These were the pitiful effects of rabies which, once set in its course, left those who watched powerless to help.

In France the disease never gave rise to more than a few hundred cases each year, but its effects were so spectacularly awful that officially backed attempts had been made to find an antidote. There had been many quack remedies which had produced only death and distress. The variations on the only known reliable method, cauterisation, could themselves be fatal; they included sprinkling gunpowder over the wound and setting a match to it, and putting nitric acid directly on to the open flesh.

And so when Pasteur began to work on rabies the problem was a huge challenge, every attempt in the past having failed to throw any light on the nature of the disease. But besides the challenge, there was a real danger. It was known that rabies was spread by the saliva of mad dogs and other animals, and it was in saliva that Pasteur began to look for a microbe. An accident in the process of the handling of their material could have meant an unpleasant death for any one of the team involved. It was part of Louis Thuillier's ironic life history that he survived some of the early work on rabies with Roux and Chamberland, only to die of cholera in Egypt.

Emile Roux, the doctor in Pasteur's team, his great experimental abilities well recognised by the Master, carried out some of the most important early work seeking the microbe. But none of those who set out to investigate rabies actually saw a micro-organism, even though Pasteur, rightly, was convinced that one was involved. The thing he succeeded in injecting into a rabbit, in order to infect it with the agent of the disease present in the saliva of a child who had died from it, was much smaller than anything he had yet dealt with. The micro-organism responsible for rabies was what is now called a virus, invisible under the ordinary microscope, and one which Pasteur was unable to grow in the way he had other microbes. But in spite of not being able to recognise what it was he was chasing, his aim nevertheless was to use it to make a vaccine.

Growing and studying the micro-organism was work which would take three years to complete. The important initial assumption which Pasteur made was that, since this unseen microbe attacks the central nervous system, that was where it must be grown. The medium he chose was the brain and spine of living rabbits. The technique that would be developed – brilliant and original – would bring Pasteur, Roux and the others into head-on conflict with anti-vivisectionists.

Emile Roux was a highly intelligent man. Ascetic, as piercingly observant and acid-tongued as his Master, he was less prone to the intuitive gamble. He began his work with the simple aim of seeing how long the rabies virus remained active in the spine after an animal was dead. In order to dry the spine, he had designed an elegant little flask in which the tissue could be hung, with two entrances so that air could flow freely in and out. This flask was to produce an unpredictable conflict in Pasteur's team itself.

Pasteur's new assistant was his young nephew, Adrien Loir. One day as uncle and nephew were passing through the room in which Roux stored his samples, Loir turned to see Pasteur holding one of Roux' newly designed flasks. 'Who put this there?' said Pasteur. 'It could only be Monsieur Roux. It's his shelf space,' the young man answered. Loir watched Pasteur go into the corridor, hold the flask up to the light, then continue to stare at it in silence for several minutes. Pasteur returned to the room, replaced the flask on the shelf from which he had taken it, and said nothing. He then immediately headed back to his own laboratory where he ordered his glass-blower to make a dozen flasks, identical in design to Roux', but larger.

Pasteur had conceived the idea of weakening or attenuating the rabies virus by suspending and drying the spinal cord of rabbit over potash in one of Roux' flasks, thus allowing oxygen of the air to penetrate the cord. Soon Pasteur was having his nephew experiment with the new flasks. At first Loir was clumsy, burning his fingers on the potash as he tried to suspend sections of rabbit spines on pieces of cotton in the flasks. But by next day he had three satisfactory preparations sitting in the balance room. That evening Loir was called into the same room by Roux who was clearly subduing intense anger.

'Who put those three flasks there?'

'Monsieur Pasteur,' Loir replied.

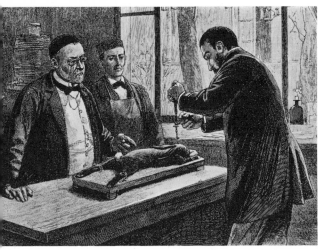

ABOVE LEFT: Pasteur experimenting on a chloroformed rabbit in the search for the cure for rabies

ABOVE RIGHT: Emile Roux

BELOW: Pasteur in his laboratory during work on rabies in 1884

'He went into the incubator?'
'Yes.'
'He saw the flask on my shelf?'
'Yes.'

Roux said not another word. From that moment experimental work on rabies was to be no more concern of his. Pasteur had over-stepped the master-pupil boundary. Roux believed himself plagiarised.

Pasteur was now forced to continue his work without the undeniable talents of Roux. The method eventually developed in the laboratory was to let the rabbit spine dry and the virus grow progressively weaker. Then, in a fourteen-day-old spine, Pasteur was able to show that the virus was almost inactive. When it was crushed and injected into a healthy dog, it no longer killed. He now continued to inject the same dog with material from a spine that had been dried for only thirteen days, then one dried for twelve days, and so on until the animal could resist the fresh, virulent rabies virus. He had made the dog immune to rabies. And he was able to repeat his success with fifty dogs.

Again, Pasteur had no idea how his vaccine worked. But this did not prevent him from having such sweeping confidence that he could think the time might have come when he could try out his results in human beings. As with so many examples of Pasteur's sweeping confidence it was not universally shared. But more important, it was not even shared by those in his own laboratory. Some of the young men whose scientific reputations had grown up under his enormous shadow, and who had followed the ways of the Master with such religious devotion, were beginning to mature and to hold strong views in the same way that their teacher held strong views. Emile Roux, now less uncritically devoted than he had once been, was one of these. He seriously doubted that the time had come to put human life at risk. Should the method fail, and should a patient actually catch the disease from the vaccine, the consequences for that patient would be truly terrible. No matter what spurred Roux' scepticism, there could be no doubt that his logic was sound.

Pasteur too had his doubts, but circumstances suddenly forced a decision. Rumour had spread that he was working on a cure for rabies. It had even spread as far as Alsace. It was from that region that, on 6 July 1885, a young mother made the long journey to Paris with her nine-year-old son. The boy was covered in bites. Joseph Meister had been on his way to school when a mad dog had set upon him. Luckily a brick-layer, seeing the attack, had beaten off the animal with an iron bar. This, however, was only after the dog, slavering at the mouth, had sunk its teeth into the boy fourteen times. Now Pasteur stood and looked at the wounds, having to weigh the odds of the child dying from rabies, or from his untested vaccine.

Pasteur was not in an enviable position. He was not a doctor. But even if he were, there was nothing in a doctor's code of practice to support him. The unhelpful Hippocratic Oath, for example, could have told him nothing about whether what he was thinking of doing would be for the good, or for the ill of the patient. All he could be sure of was that he was carrying out an operation which was unquestionably experimental: and that the Oath frowned on. From a scientific standpoint his position was weak. He had not even succeeded in following his own advice of 'hunt the microbe'. He had not managed to identify the germ responsible in an animal. And quite apart from the fact that he, the over-sensitive, emotional father of three dead children, might end up with the death of a child on his conscience, the publicity – which he desperately cared about – would be disastrous.

But in front of him was an imploring mother and the pathetic child who stood a good chance of dying an ugly death unless something was done quickly. Pasteur knew that there were many in the leading medical circles of Paris who were strongly opposed to what he was contemplating. Worse there were some in his own laboratory who had the severest reservations, including Roux: the uncompromising Roux who would later say to a woman who offered to pay for rabies treatment, 'I can assure you, dear Madame, that all my services are free, including autopsies.'

Pasteur decided to share his burden with two colleagues from the Académie de Médecine, Edmé Vulpian, a physiologist, and Jacques Grancher, a physician. They looked at the deep bites on the boy's hands and both men told him what he wanted them to tell him: that they believed that Joseph Meister would die from rabies if not treated. Even more persuasively, they said that Pasteur's experiments on dogs justified an experiment on a human.

Pasteur lost no time and, under Grancher's medical supervision, immediately had the boy injected in his side with the first dose of vaccine. The theory on which Pasteur was banking his success was that rabies develops slowly in humans, as it does in dogs, and that, after infection, there was a long enough incubation period for the vaccine to be effective. Already it was two and a half days since the dog had sunk its teeth into Joseph Meister. There was no way of knowing whether the safe period, if it existed, had not passed.

Over the next ten days Pasteur planned to have twelve inoculations, 97 each successive one made up from a more virulent piece of rabbit spinal cord. Grancher would administer each one to the child. Pasteur won Joseph's confidence and, at first, the boy reacted well. But there were upsetting moments in the days ahead. Pasteur wrote to his son: 'There is some reaction, which is becoming more intense as we approach the final inoculation, which will take place on Thursday, 16 July. The lad is very well this morning . . . He had a slight hysterical attack yesterday.'

Joseph Meister was not the only one to suffer traumatic side-effects. Pasteur too was badly affected by the strain. Throughout this period he slept badly at nights, suffered from nightmares, and had acute anxiety over what might be the result. Then an incident occurred which succeeded in heightening even the existing tension. It was the simple error of a moment's inattention: the ghastly slip of the hand that could have ended in the most appalling tragedy. Grancher took the needle to prepare one of the injections, then moved with it towards the boy. Inconceivably, he fumbled and plunged the metal point into his own leg. He had inoculated himself with the virulent rabies he was attempting to kill in the child.

Now, as an act of necessity, and not merely one of faith in his own judgement, Grancher had to make the decision to submit himself to the same treatment that he was at that moment administering for the first time to a human subject. If he did so it would make him the second human guinea-pig for rabies vaccination before the first survived the course.

Adrien Loir and Eugène Viala had prepared the rabbit spines for Grancher's injections. They were the two young assistants present in Pasteur's room when, in utmost secrecy, Grancher prepared himself for the inoculations. They had both listened to Emile Roux' harsh criticisms of the techniques they were involved with and, by now, such was both his powerful personality and logic, partly shared his fears of a disaster. Loir had even gone so far as to try to persuade Grancher not to risk his life. Dryly Grancher had replied, 'Do you think, young man, that I would do this job every morning if I wasn't absolutely sure of the method?' As a consequence of his confidence, Loir and Viala upended their own logic and decided since they could not beat him they would join him. They too elected to be inoculated.

Paradoxically it was young Loir to whom fell the perilous honour of injecting Grancher. However, before he could do so Pasteur intervened. 'You must first inoculate me,' he said to Grancher. But Grancher, arguing the idiocy of making a risky inoculation of a man who had not been exposed to rabies, and emphasising the wisdom of protecting one who had, absolutely refused the old man's order. Pasteur's annoyance at his temporary loss of command was increased further by Grancher's agreement that Loir and Viala, who had been in daily contact with the rabies microbe, should be injected. There then followed the extraordinary series of events when, closeted in the little room, each inoculated the other. It was the first of fourteen such operations to which all three had to submit.

However, the extraordinary tale does not end there. The clandestine group was not, as it believed, unobserved. Emile Roux had got wind of some crisis in the rabies laboratory and had actually succeeded in seeing something of what was going on by putting his eye to the keyhole of Pasteur's room. And, in spite of the far from admirable way in which he had come by his information, Roux had no hesitation in revealing his worst suspicions. A few days later, finding Loir alone, Roux grabbed him by the arm and demanded to know what they had been up to that day in Pasteur's room. Roux believed that he had seen Pasteur himself being inoculated and Loir had to reassure him that this was not the case.

By now, of course, as Roux well realised, it was too dangerous for the three men to stop administering the treatment to each other. There was no alternative but to continue the full fourteen days' course of inoculations. Roux, who by this stage of the conversation was becoming hostile, hinted maliciously that Loir's parents ought to know what was going on; Pasteur, after all, was Loir's uncle. Loir desperately feared that the worry would drive his oversensitive parents to distraction if they discovered the truth, and were told of the real nature of the risks involved. His interrogation by Roux ended with the older man snatching up his hat and storming from the room. Nevertheless, in spite of his fury, Roux never mentioned what he had discovered to Loir's parents. That would be left to Pasteur's thoughtlessness. He would let slip the truth two months later at a lunch party. In spite of frantic efforts by Loir and Madame Pasteur to steer him off the conversation, Pasteur would freeze the party by revealing all, leaving Madame Loir aghast and on the verge of a breakdown.

It is a remarkable fact that from this indescribable muddle of jealousies, insensitivities, paradoxes and plain error – but involving highly intelligent human beings in scientific creativity and in the creative application of science – there emerged something precious and inimitable: the seeds of work which would grow into a lasting, civilising contribution to man's way of life.

After the end of the fourteen-day period of inoculations, a period of fearful uncertainty set in. Joseph Meister was sent back to Alsace on 27

July. Pasteur took an enforced rest in Burgundy with his daughter, worrying over the multiple consequences of what he had done. By 3 August it was becoming clear that all might be well: 'Very good news last night of the bitten lad,' wrote Pasteur. 'It will be thirty-one days tomorrow since he was bitten.' He had survived with only his bite scars to show for the terrible experience. Grancher, Loir and Viala too lived to tell the tale, although the true facts of Grancher's near-fatal error and its repercussions were hidden for many years.

The news of success with Meister spread throughout the nation. Again Pasteur's was a publicised success story. By the middle of October he was supervising the injections on a second victim of a rabies-mad dog. And on the 26th of that month he was sitting in the Académie des Sciences listening to Vulpian who, on the basis of having seen one cure only, was eulogising on the infallibility of Pasteur's treatment. 'I say infallibly,' he said, 'because, after what I have seen in M. Pasteur's laboratory, I do not doubt the constant success of this treatment when it is put into full practice a few days only after a rabid bite.' Not all his audience was so convinced of Pasteur's God-like facility.

It was about this time that a most singular figure drifted – or perhaps one should say shuffled backwards – on to the central scientific scene. It speaks wonders for the conservatism of popular imagery that the common picture of a scientist as a predictable introvert has never been shaken by the reality of Elie Metchnikoff. This exceptional man, in spite of a life of wild unpredictability and great achievement, has sunk into an inexplicable obscurity, but from which one day he will undoubtedly levitate himself.

Metchnikoff's early years could well have provided the material for a Chekhov play. The youngest son in a land-owning Russian family, he saw his mother's inheritance frittered away by his father. The mother considered that her weakly child had sensibilities which were too feminine to allow him to study medicine; for this reason he had to reject it in favour of zoology. He was a highly temperamental youth and frequently threw himself into a frenzy when his studies were disturbed by anything as apparently ordinary as a dog's bark or a cat's mew. As an adult he became a manic depressive: a condition he observed with the same fascination as he did any other of his own physiological states. Even when he suffered a serious heart attack he took detailed notes of its effects on himself. And when he attempted to commit suicide for the second time, by injecting himself with relapsing fever, he was interested to discover whether the disease could be inoculated through the blood. The answer, though in his case not fatally so, was in the affirmative.

On the first occasion, in Geneva, in one of his fits of depression, this time caused by his fear of losing his sight, he gave himself morphia. Having taken too strong a dose, he was violently sick and threw up most of the poison. He then took a hot bath and tried to finish himself off with a resounding fever by exposing himself to the cold on the Rhône bridge. But standing there shivering, he saw a great cloud of insects flying by, which he identified as the short-lived mayfly. And as he stood gazing at them, he asked himself how could Darwin's theory of natural selection apply to these ephemeral things? In the few hours the creatures have to live, he asked, how can they possibly adapt themselves to their surroundings? And the marvel and argument of science, at least so he told his second wife, re-established his link with life.

This wife, Olga, was married to him in her early teens. He had chosen her to be his 'helpmeet' as he called it, when he was 30. One of the reasons he married a girl so young, quite apart from the fact that he had an almost uncontrollable affection for little girls, was in order to satisfy a theory: in this case a theory of marriage. He firmly believed, and he drew on statistics of suicides to confirm his point, that one of the reasons for disharmonies and unhappiness in civilised societies was that although people were marrying later, the age of puberty remained the same. He thought that the long gap between puberty and marriage was the cause of contemporary society's malaises. His marriage to Olga was made, like all his actions, with the best of motives; and in this one case at least, it was a marriage of stability: she was with him at his death.

At the peak of his manic moods he was at his most attractive and creative: Olga was fond of comparing his demeanour, his pale bearded face and his saintly manners, to those of Jesus Christ. And the ideas he plucked as though from nowhere, whether they were concerned with marriage, ageing, natural selection, or directly with his own subject, though many of them were plainly naïve, had the most appealing simplicity and originality.

In 1882, thanks to an inheritance acquired from his wife's parents, Metchnikoff was carrying out some research on zoological specimens in the Mediterranean. He had rented a small apartment overlooking the Straits of Messina, in which he had settled his wife and the rest of her inheritance – her five younger brothers and sisters. It was an ideally beautiful place in which he could pull himself out of a great period of depression which had followed his forced resignation from the University of Odessa because of his radical political views.

One day, when his spirits were on the rise, being left in peace by the whole of the family who had gone off to watch a circus, he sat at his microscope and began watching some mobile cells in the transparent larva of a starfish. And then:

A new thought suddenly flashed across my brain. It struck me that similar cells might serve in the defence of the organism against intruders. Feeling that there was in this something of surpassing interest, I felt so excited that I began striding up and down the room and even went to the seashore in order to collect my thoughts.

This picturesque, but apparently simplistic view of a body being manned by defenders who warded off attackers was typical of Metchnikoff. But it was not naïve. It was the first real step in explaining the mechanism by which a human organism could be, or could be made to be immune to a disease. Metchnikoff immediately worked out in his mind a simple experiment by which he could test his theory. From the garden of the apartment where they were living and where, a few days earlier, the children had been using a tangerine bush to decorate as a Christmas tree, Metchnikoff took some rose thorns and carefully pushed them under the skin of some beautiful specimens of starfish larvae which he had collected.

He knew it would take several hours to see any result. Now at one of his manic peaks, he spent the whole of that night in an excited state, getting no sleep at all, until he was able to look down his microscope in the early hours of the morning. What he saw confirmed his theory precisely. Just as mobile cells surround a splinter in a man's finger, so too they had surrounded the rose thorn in the simple starfish larva.

Metchnikoff was now in a position to form a theory of how a microbe, intruding into an organism such as the human body, is dealt with. He reasoned that, since microbes cause inflammation – just as a splinter of wood causes inflammation – mobile cells must be called into action to destroy the microbes by digesting them. He was able to test his theory precisely and beautifully by inoculating transparent starfish larvae with microbes and watching them being surrounded by cells. He could then see how close to fact was his childlike idea of the body as a battlefield, fought over by warring germs and protective cells.

In Messina at the time were two distinguished Germans: Kleinenberg, the Professor of Zoology at Messina; and Virchow, the great pathologist who had so crushed the young Koch. But whatever it was in Koch that had caused Virchow to bristle was not present in Metchnikoff when the grand old man of German science unexpectedly paid a call on the young Russian to look at his work. Both Kleinenberg and Virchow were encouraging, though Virchow pointed out that Metchnikoff's theory was opposed to the then current theories of the cause of inflammation. It was believed that the leucocyte cells in human blood, rather than eating up microbes, as Metchnikoff was suggesting, made a suitable medium in which the microbes could grow.

But Metchnikoff had confidence in his armies of leucocytes (*phagocytes* was the name he coined for them) though as yet he had never clapped eyes on one. However, he had every intention of doing so. He took the transparent waterflea, *Daphnia*, whose blood cells he could watch under his microscope; then he showed how, when it was diseased with a microbe – in this case the spores of a vegetable parasite – his phagocytes rushed to the flea's defence. With enormous enjoyment he wrote up his work as though he were an observer on a hillside watching a squadron of enemy light cavalry being outflanked by the superior heavy dragoons of the home force. The phagocytes were, of course, victorious.

Metchnikoff now had to ask himself the question, why do phagocytes in some animals and humans succeed in eating up invading microbes, and why do some fail, so leaving the body a prey to disease? What, he was asking, is the mechanism of *immunity*? It was a vast question: one of desperate importance to Pasteur who, without being able to answer it, was at that time producing some startling and hopeful results with his rabies work.

Metchnikoff began to observe rabbits infected with anthrax, some of them having been vaccinated, and some not. His conclusion was that the phagocytes in the body of a vaccinated rabbit become gradually accustomed to fighting the weakened microbes, and are then able to put up a better performance when virulent anthrax bacillae enter the rabbit.

He was elated by this theory. He believed, rightly, that he had hit on something extraordinary. Through his new friendship with Virchow, he was able to find a first-class means of publication in Virchow's own journal, *Archiv*. But the reaction to his work when it appeared in print took Metchnikoff by surprise. The significance of his elegant observations went, for the most part, unnoticed. And those who bothered to read the paper attacked it in a bitter fashion, which hurt the childlike Metchnikoff. He took all attacks on his work personally since so much of himself was involved in his science. He believed that his sort of research was most likely to lead to a unified conception of life as a whole, and that science could be used as a great social force to better the condition of man. He was genuinely astounded that others did not see the theory and the function of science as he did, just as he was shocked that others did not see that harmony in life was to be got from the stark existence he lived, shunning any luxury and rejecting any object that did not have a use.

Even in Russia, his own land, where Metchnikoff was a big fish in a small scientific pool, there was disagreement about the value of his work, as well as suspicion of his left-wing social theories. He was, he felt, a prophet without honour, and he decided to turn his back on his country. He headed again for Western Europe, and for the laboratories of the greatest workers in his field, Pasteur and Koch, to discover what informed opinion thought of him.

Pasteur cared about understanding how his rabies vaccine worked: establishing a rational understanding of the processes of science was his *raison d'être*. So too did Roux care. But Pasteur and his assistant's interpretation of the responsibilities which their new-found knowledge threw on their shoulders was vastly different. Pasteur, the pragmatist, his eye always on the main chance, having been pushed by circumstances into applying the rabies vaccine, was in favour of publishing the work without more ado. Roux, cautious, critical and sensitively aware of the fact there were problems still to be resolved, was strongly against publication which would put human lives needlessly at risk should the work be in error. In the end, Pasteur published, but without any name other than his attached to the work, in spite of Roux' contribution to it. It was the admission of the serious rift in this fruitful scientific union.

The result of the publication, and of the publicity created by Pasteur's own graphic descriptions of the disease from the floor of the Académie des Sciences, soon had patients who were suffering from mad dog and wolf bites flocking to the Left Bank of Paris in search of a cure. The demand was such that Pasteur had to create a special clinic to treat the cases, along with a laboratory given over entirely to producing the vaccine. Before long there was a queue of invalids from all parts of France outside the clinic, and inside, rows of sterilised flasks in which hung rabies-infected spinal cord – the first stage of the vaccine preparation. Soon there was to be news that patients were on their way from America and Russia to join the queue of assorted humanity, of rich and poor, some scarcely scratched and some severely mauled, all patiently waiting to be inoculated. The newspaper publicity for Pasteur's work was now at its peak. His public relations were at their most successful. But he was treading a fine line of risk. Many were aware of the fact. 'Tattooing', some sceptics called the injections; and they watched and waited.

There was not a long wait. On 9 November 1885 a girl of 10 was brought to Pasteur. She had been bitten on the head by a mountain dog on 3 October. The wound was still suppurating. Pasteur examined the wound and calculated the number of days since the bite: thirty-seven. It was plainly too late to vaccinate and he made this clear to the parents of the girl, Louise Pelletier. But the parents pressed him – as they had to. And Pasteur, with no other way open to help the child, and the chance that his method might be even more effective than he had ever believed, gambled – as he had to. He gave Louise Pelletier her first inoculation on 9 November. She returned to school after the course of injections was over, showed signs of rabies within only a few days, was shortly in convulsions, and dead by 2 December.

It was what the sceptics had waited for. Louise Pelletier, it was concluded, had been killed by the virulent rabies injected by Pasteur. He now had to face the consequences of the backwash from his own wave of publicity. The possibility that he was a manslaughterer was given as much newspaper space as the possibility of his being a life-saver. It was pointed out how relatively rare a disease was rabies, although its spectacular symptoms made it appear more common than it actually was. Many people who were bitten by mad dogs never developed any of its symptoms. How could Pasteur claim such a success rate for his vaccination when he had no clear idea of whether his patients were infected?

In medical circles the feeling that Pasteur had overstepped the mark of scientific propriety was growing. Grancher, still administering the inoculations under Pasteur's supervision, was deeply implicated. To question Pasteur's motives was to question Grancher's. One day, as Grancher was walking into the Paris Medical School, he heard a voice saying, 'Yes, Pasteur is an assassin'. He turned to see a group of his colleagues who, seeing him, broke up and went off in silence.

Pasteur realised that the case against him was strong. Neither were his attackers simply those who were jealous of his unbroken line of successes nor objected to his personal publicity. One of the men who put his work to the most stringent analysis was a distant relative, Michel Peter, Professor of Medicine at the University of Paris. The platform from which he attacked Pasteur was a familiar one: the Académie de Médecine. He pointed out just how rare a disease was rabies; he himself had seen only two cases in thirty-five years of practice. He also produced the crushing isolated statistic that in Dunkirk, in twenty-five years there had been only one death from rabies. Yet, in the same town in one year, since the application of the Pasteur method, there had again been one death. Peter, like many others, believed that Louise Pelletier was a victim of 'laboratory rabies', inflicted by Pasteur.

Pasteur could point to the statistics that gave his method support. By 1 March 1886, of 350 people treated, only one had died. And an official inquiry set up in the City of Paris estimated that the mortality rate from bites by rabid dogs was 16 per cent.

Pasteur's work, affecting human lives as it did, touched human sensitivities and reactions. The swell of public approval for his treatment rose and fell. On one great wave of enthusiasm a fund was set up to found an establishment for the treatment of rabies; it would be called the Pasteur Institute. Support came from abroad, with several European newspapers running fund-raising campaigns. But, just as quickly, emotions reversed. There were more deaths at the Pasteur clinic; they were again attributed to 'laboratory rabies'. After one boy died a lawsuit was threatened which, had it been allowed to follow its full course, could have left Pasteur branded as a manslaughterer, put a halt to the work and put the creation of the new Institute in peril.

In April 1886 a British Commission was established to look at Pasteur's claims, with a view to establishing the vaccination procedure in the British Isles. The Commission looked into the treatment in detail and vindicated it. But as Michel Peter pointed out when the findings were published, the Report did not recommend 'the establishment of a Pasteur Institute in London, but instead recommends a more rigorous enactment of police regulations on dogs as a means of rabies prevention'. The average number of deaths annually from rabies in Britain at that time was a mere forty-three.

It was true that, using a pure economic reckoning, the cost of saving a few French lives would have been far less had the money been used to mount a publicity campaign to have all dogs muzzled. But these were the arguments of a mind with a limited vision. There can be no doubt that Pasteur played dice when he risked treating his patients, but the long-term stakes were far greater than those on view at the time. It has to be conceded, however, that this argument is difficult to sustain when the gamble is with human lives. But what Pasteur had done in his work was to pave the way for an investigation of methods of immunity from other killer diseases, some of them infinitely more drastic in their effects, than rabies.

One of the men who saw the importance of Pasteur's vision was Emile

107

Roux, a stranger for some time now to Pasteur's laboratory. When the attacks were strongest on Pasteur's reputation, Roux came to the defence. He put an enormous amount of skill and energy into testing for the presence of rabies in dogs in order to quell the criticism that treatment was being given to uninfected patients.

As it never failed to do throughout his life, Pasteur's public image began to rise against the snapping of criticism. The defence of this particular work, however, was longer and more exhausting than most. Bit by bit events moved in his favour. Support for the Pasteur Institute mounted. At one theatre benefit, Gounod conducted his *Ave Maria* and, 'in an impulse of heartfelt enthusiasm, kissed both his hands to the *savant*'. At the Académie de Médecine opinion began to swing in Pasteur's favour and he could now quote more favourable statistics. From 1880 to 1885 sixty cases of deaths from rabies had been reported in Paris hospitals. From November 1885 to August 1886, during which time rabies vaccination was in operation, there had been only three deaths, and two of these victims had not been vaccinated.

The official reinstatement of Pasteur's reputation came in July 1887 when he was elected Life Secretary of the Académie des Sciences. But this period had taken an enormous toll on him; physically and mentally it had been the most trying of his life. As well as having been deeply involved in more than two years of medical activity, with patients daily clamouring at the door of his laboratory, he had had to evaluate the pathetic claims of these same patients and, if his method was at fault, bear their deaths, when they occurred, on his conscience.

In October 1887 he had his second paralytic attack. As he sat writing a letter in his room, he tried to speak to his wife, but she saw that he had lost the use of his tongue. He insisted on lunching with his daughter, as he had planned, and tried, as he had on the first occasion, to fight off the symptoms. But a week later he was struck down again. He was to live on for several years, but his life of phenomenal creativity was at an end.

CHAPTER EIGHT

Metchnikoff spent several weeks of 1887 in Western Europe, hopping about in his peripatetic fashion from laboratory to laboratory, painfully bearing his breast to the many who had chosen to attack his theory of phagocytosis. At last he arrived in the Berlin laboratory of Robert Koch where, anxious to please as always, Metchnikoff spread out some specimens for the great man to look at. But Koch, in spite of his experience as a young doctor trying to make his way in pure research, was as crushing to the Russian as others had once been to him. Metchnikoff was bitterly hurt, and was grieved even more by Koch's assistants, who lined up behind the Master waiting to see on which side his axe would fall before they committed themselves to the rejection of Metchnikoff's beloved phagocytes.

Koch never knew it, but this was a critical visit which would determine how Metchnikoff would spend the rest of his life. Metchnikoff had just come from Paris. His reception by Pasteur there could not have been in greater contrast. Nervously he had sat in the laboratory looking at the 'old man, rather undersized, with a left hemiplegia . . . His pale and sickly complexion and tired look betokened a man who was not likely to live many more years'. Pasteur, nevertheless, was not too decrepit to know that Metchnikoff desperately wanted to talk about microbes, and in particular about microbes related to phagocytes. He sensed that, at the core of this ascetic and oddly mannered young Russian's theory there was a visionary light. He was encouraging. It is scarcely surprising, therefore, that within a few months Metchnikoff was back in Paris working in Pasteur's laboratory. Metchnikoff established himself in Paris and became a fixture of the laboratory, much loved and admired in spite of all his eccentricities. When the laboratory was in financial difficulties, Metchnikoff refused his salary. He was never to leave the place except, as he himself wished it, when heading for Père Lachaise Cemetery.

One of Pasteur's assistants who approved wholeheartedly of the Russian's spartan life-style was Emile Roux. Metchnikoff found his personality attractive. Temperamentally, however, the collected, handsome Roux was very different from the disorganised Metchnikoff: his attitude to women, for example. Said Roux, 'Women are like drugs. When they no longer act, one must change'. (And one of his considerable number of female friendships was with Madame Metchnikoff, now no longer in her teens.) Their work too, though moving towards the same end, was based on fundamentally different ideas.

Roux, along with Alexandre Yersin, had become deeply involved in

the problem of immunity. Yersin had spent a short time in Berlin working with Koch and had brought back to Paris a detailed knowledge of
what the Germans were up to. Löffler, Koch's assistant, had just succeeded in isolating the bacillus responsible for diphtheria – small,
needle-like rods. As the controversy over the rabies vaccine was raging,
Roux and Yersin had been trying, without success, to find a similar
vaccine for diphtheria.

Roux took his microbe samples for his inquiry from children at two
Paris hospitals, the Enfants Malades and the Hôpital Trousseau. The
death-rate from diphtheria in these places at that time was about 50 per
cent. Still the only treatment for most of the children with throat mem-
branes affected by diphtheria was to have a tube pushed into the neck to
allow them to breathe. The diphtheria bacillus, Löffler had shown,
remained localised in the throat of its victim. Yet the whole body was
affected as though it were poisoned. Roux had set himself the task of
finding out why. Eventually he was able to cultivate microbes, let them
incubate – as they did in a child's throat – for several weeks, then filter
off the liquid produced. He found that this liquid was a deadly poison,
an ounce of which could dispatch 75,000 dogs. In diphtheria the microbes did not kill, but this substance which they secreted did. Roux
believed that it should be possible to use this toxin as a vaccine. The
theory was apparently in direct opposition to that of Metchnikoff, who
believed that immunity to a disease could not be given by some inanimate poison in the blood, but by his lively cells, the battling phagocytes.

But whatever else these differences of views revealed they showed no
rift in the generally smooth relations between the workers in the Pasteur
Institute. Apart from a tricky period of antagonism between Pasteur and
Roux during the rabies work, this laboratory had all the appearances of
being one where a common aim was shared. Here, whatever were the
individual motives in the search for scientific truth, they produced
remarkably few personal bitternesses and petty jealousies.

The same cannot be said of the other leading bacteriological laboratory in Europe, led by Robert Koch. The achievements of the German
school had been enormous. Pasteur's success both with his rabies vaccination and his public relations had temporarily redirected the lime-
light to France. But the intense rivalry between the two laboratories
remained, and the one in Berlin was poised to make some enormous
advances in the understanding of the nature of diseases. The workers
who were to make them, however, were not to come out of the experience unscarred by mutually inflicted wounds.

In the few years before and after 1890 Koch's pupils were to identify the microbes that caused typhoid fever, diphtheria, erysipelas, tetanus, glanders and some forms of pneumonia and meningitis. And the identification of the cause of a disease, as by now even the general public was aware, was the first step towards conquering it. Justifiably, this work gave Koch a phenomenal reputation. But in scientific circles the names of his assistants were also becoming well known: men such as Gaffky and Löffler and, more recently, Emil Behring, the energetic ex-Prussian army doctor, Shibasaburo Kitasato, one of several Japanese taken on in German laboratories about this time since they added little to the laboratory's budget, and Paul Ehrlich, an untidy chain-smoker of cigars who adored his laboratory work and was willing to work without pay if necessary.

When Behring arrived in Koch's laboratory the wall of protective secrecy which Koch had built up in order to safeguard the priority of his work was securely established. The spirit of competition worried Behring not at all. He enjoyed a race just as he enjoyed a good row; his military career had given him the stamina for some of the infighting that he was to take part in during the next few years. He was ruthless and he knew how to use his friends to his own advantage: characteristics which did not endear him to all his co-workers.

Behring, like Koch, was one of thirteen in a poor family. But in his youth he was less of a dreamer than his professor had been. Behring was a pragmatist. He prepared himself for a priesthood since this was a means by which he could have free university training, transferring later to the army medical school. Once in the army, and with theological thoughts behind him, the young Behring ran up enough gambling debts to cause a minor scandal before, by contrast, he developed a more consuming interest in infectious diseases. He was 35 when he joined Koch at the Institute of Hygiene in 1889.

Within a short time Behring had become interested in the diphtheria problem and the method of producing toxin which Roux and Yersin had turned into public knowledge. However, it was regrettably typical of the laboratory in which Behring was now working that, at that time, there was at least one other man occupying a nearby bench trying to turn up a diphtheria vaccine. The spirit of the place was such that each was moving entirely independently of the other. If he consulted this other man, Karl Fränkel, Behring recorded no debt owing to him.

Behring's approach to his new problem was guided by his army experience. There he had become very familiar with the use of iodoform as an antiseptic wound dressing; the efficient way in which it acted prompted him to try out iodoform on some of the toxic products of certain bacteria. He found it very effective as a way of reducing their toxicity.

He now began to investigate the effect of treating cultures of diphtheria bacilli with another iodine compound, iodine trichloride, with which he then inoculated a number of guinea-pigs. Some survived and, significantly, he found that they were now immune to diphtheria. He was also able to show that minute amounts of diphtheria toxin immunised the animals. By taking blood serum from an immunised guinea-pig and injecting it into another animal he now found he was able to transfer the protection to the second guinea-pig. Not only that, an animal already suffering from diphtheria toxin could sometimes actually be cured. Although the serum did not affect diphtheria bacilli, it neutralised the toxin they produced. This was the discovery of an *antitoxin*. It was an epochal piece of research. For Behring it was worth a permanent and deserved place in the history of medicine. Behring would make good use of the place.

If interpersonal relationships in general were bad in Koch's laboratory, nevertheless one fruitful, if unlikely, union had been made: that between the extravert ex-army officer, Behring, and the polite little Japanese, Kitasato. Only the year before they met in Koch's laboratory, Kitasato had identified the microbe which caused tetanus and shown how, like the diphtheria bacillus, its danger to man lay in the toxin it produced. Together he and Behring were able to show that a rabbit could be immunised against tetanus in the same way as it could against diphtheria. The use of the rabbit meant that Behring and Kitasato were able to collect relatively large amounts of immune rabbit blood serum. Enough, at least, not only to protect a mouse from a lethal dose of tetanus toxin, but to cure one already suffering from tetanus.

Antitoxin was a wonderful discovery since immediately it opened up the possibility of finding a vaccine for diseases such as diphtheria, which took so many children's lives. Behring consulted others in the laboratory besides Kitasato, who clearly deserved whatever acknowledgement he got for his contribution to the joint work. The young Paul Ehrlich was undoubtedly one who made suggestions to which Behring not only listened with attention but was more than glad to adopt. Naturally, Behring would, in his situation, be expected to consult Koch. He did and, it seems, Koch counselled caution before publication. Behring, after all, had no idea what this antitoxin was. It was a mystery substance. The name antitoxin was convenient, but it explained nothing.

Behring, however, had caught on to Koch's idea that a publishable piece of scientific work was a property to be safeguarded. On 4 December 1890 Behring and Kitasato published their results for tetanus. A week later Behring published a second paper, which carried his name alone, and which described his experiments with diphtheria. Behring had made his mark.

Koch himself was going through a difficult period. For six years now he had produced nothing that could match the originality of his early scientific successes. Inevitably, as he grew older, his habit of working in secrecy was turning him into a nervously suspicious individual who saw a conspiracy behind every polite question. Soon he was to complain to a friend, 'Whenever I start something new, a swarm of ill-wishers and self-seekers batten on to me'.

Domestically, too, he was not a happy man. Twenty years and more of marriage to Emmy had not strengthened their bonds; the financial worries that tied them together had disappeared with Koch's increasing fame. At the age of 45, bespectacled, bearded and with thinning hair, Koch began to look around him for something Emmy could not offer. His eyes lighted on a surprising subject. He was sitting in an artist's studio at the time having an official portrait painted. On one of the canvases in front of him rested the portrait of a seventeen-year-old girl. A few enquiries told Koch that she was a student of the artist, she had played small parts at the local Schiller theatre and she was the illegitimate daughter of a Berlin workman; she was also, as Koch could see, unquestionably pretty. A meeting was arranged, and before long the girl, Hedwig Freiburg, was the centre of Koch's life and thought outside the laboratory.

This crisis in Koch's domestic situation also happened at a critical time in his scientific affairs. How much one predicament influenced the other is difficult to guess. Although it was well known that he was working on tuberculosis he hid the details of what was going on behind his laboratory doors with even more caution than usual. The German Government too was well informed of Koch's preoccupation. His successes in the past had brought nothing but honour to Germany, and now his fame was a commodity that would be exploited in the cause of national prestige. It so happened that the Tenth International Medical Congress was to be held in the summer of 1890 in Berlin. If, as it was rumoured, Koch was on the point of making some phenomenal breakthrough in the understanding of tuberculosis, the timing would be magnificent. A cure for an unconquered scourge announced by the discoverer himself in the nation's capital before one of the most distinguished international medical gatherings ever held would be a shining demonstration of national intellectual supremacy.

The situation was made quite clear to Koch by Gustav von Gossler, the Minister of Culture. And Koch was led to believe that not just the nation, but the Kaiser himself, looked forward to Koch using the Congress as a platform to make some climactic announcement about the disease. There seems little doubt that Koch was temperamentally in favour of the patriotic and nationalistic motives urging him on, and wanted nothing more than to respond in the same spirit that the requests were made. Regrettably, as the Minister of Culture could not understand, but as Koch knew too well, neither bacteriological research, nor natural scientific research of any kind, can be made to fit into a timetable. As the date of the Congress came closer Koch, billed to speak at the opening session, with an audience expecting a sensational announcement, was so sensitive to the dangers of the trap he had helped prepare for himself that he was considering abandoning his speech altogether.

He did not, however. He stood that day in an auditorium decorated in the brash style which the Ministry of Culture had decided was appropriate to a major medical occasion. Surrounding the hesitant, physically insignificant middle-aged man, shuffling his specimens on the raised dais, were the ludicrous cardboard and plaster trimmings of a simulated temple guarded by a statue of Aesculapius, the god of medicine. Present in the hall was the scientific cream of the 7000 delegates to the conference, all of them primed to expect that the nub of Koch's speech would be something of enormous scientific and medical interest.

His introductory words made it clear that he intended to use the occasion to review the latest developments in bacteriology, all very familiar to his audience. Not surprisingly, the murmurs of disappointment turned to impatience, and soon to boredom – until he finally arrived at the words, 'I have tested a large number of substances to see what influence they would exert on the tubercle bacilli . . .'

The hall was now in silence. This was the preamble to what they had come to hear. Koch went on, cautiously listing the difficulties and failures he had encountered in many trials, then added, '. . . and I have at last hit upon a substance that has the power of preventing the growth of tubercle bacilli not only in the test-tube, but in the body of an animal'.

Koch had produced an enormous rabbit from his hat. The excitement and response in the Hall was overwhelming. Koch had done all that was expected of him as a scientist and patriot – or so it seemed. In the hubbub of the reaction few could, or cared to, listen to the qualifications to his remarks which he now began to list. His researches were not yet completed. His work so far had been only on laboratory animals. He was only at the beginning of trials which one day might lead from guinea-pigs to humans.

It was too late. By next day newspapers had announced a cure for tuberculosis and Koch was the scientific saviour whose skills, as predicted, had risen to the magnificence of the occasion. The adulation, however, did not stretch to the informed members of his audience, which included men of the calibre of Virchow, Ehrlich, Roux and Metchnikoff. What was it Koch had done? He had not actually disclosed any details. The only information he had given was that in his possession there was a liquid capable of wiping out tubercle bacilli: a magic potion the sceptics might, and did, call it. Koch would disclose details in his own good time. And for many that was proof enough. Koch's reputation was, after all, based on an unbroken series of spectacular successes. He had never failed in the past.

It was to be called *tuberculin*. Within a few months the pressure of the public reaction had forced Koch into human trials. To his credit, the first subject was himself. He survived several hours of unpleasant reaction to the injection, including pain in the limbs, difficulty in breathing and fever. At last, by November 1890, he was able to put in print the results of his eagerly awaited trials on tubercular patients. Again Koch couched his words in cautious phrases, but again the only message that was popularly extracted from his paper was that of success, no matter how limited.

The response among tubercular patients and their families was devastating. The sick and their guardians swarmed into Berlin for treatment. Tuberculin could not be prepared nor administered fast enough to meet the pathetic demand. Its price became inflated and, inevitably, a black-market in the stuff sprang up. It appeared to matter not in the least to some of the city's most reputable practitioners of medicine that they did not know what the substance was they were administering. The name of Koch attached to it was the imprimatur which guaranteed its efficacy.

It took a relatively short time to realise that the panacea was not all it seemed. All kinds of tubercular conditions were being treated with it, and some patients, it was true, at first showed startling improvements. Disaster, therefore, was all the more poignant when a relapse set in and the patient went into rapid decline. Even Joseph Lister was prepared to adopt an unquestioning faith in the elixir. He turned up in Berlin with his suffering niece, watching over her daily after Koch had personally supervised the treatment. The girl went into decline like so many of the

Inoculation against tuberculosis using Koch's treatment

others, and neither Lister nor Koch were able to prevent the inevitable tragedy.

The fiasco had a long way to run. Hotels and hospitals in the city were crowded to overflowing and injections were being administered in hotel rooms. Racketeering in the elixir became so outrageous that von Gossler, the Minister of Culture, was forced to intervene to fix the price of tuberculin. In America the asking price for a gramme was one thousand dollars. As the crown for the achievement Koch accepted from his Kaiser the Order of the Red Eagle.

By now, Koch himself was like some plaster Aesculapius, presiding impotently over premature success celebrations being played out before him. His fall was imminent, and the instrument that would destroy not only Koch's hopes, but those of many thousands upon thousands of tuberculosis sufferers, was held by Rudolf Virchow. The sceptical pathologist, by January 1891, produced evidence to show that the corpses of twenty-one cases of patients, all of whom had been treated in recent months with tuberculin, were riddled with the worst form of the disease: miliary tuberculosis.

Virchow was too distinguished in his field for his statements to be passed over lightly. He had made sure that his words were taken seriously by flatly stating that the cases he had seen were worse than anything he had ever come across in his career as a pathologist. Koch's reputation was on the block. There was uproar in Berlin medical circles. There was now no alternative to Koch putting an end to the mystery he had created and telling the world what tuberculin was.

When at last Koch broke the news to the scientific community there was a sense in many of not merely having been let down, but of betrayal. His tuberculin was a simple thing which involved the discovery of no fundamentally new principle. He had merely followed the lead of Roux who had separated the poison secreted by diphtheria microbes from the microbe itself. In the same way Koch had taken tubercle bacilli, cultured them on a glycerine broth for several weeks, killed the bacilli by heat, and filtered off his fluid. He had then tried to use this to kill the tuberculosis germs in the body.

He now stood exposed as the worst possible form of charlatan: the cost of human lives could be laid on his conscience. It was true, however, that he had never done anything other than counsel caution in both the interpretation and the application of his work. But this does not excuse Koch, nor even those who used tuberculin, from blame for the pitiful results. Had he not wrapped his experimental work in secrecy in order to safeguard his own priority, the shortcomings of tuberculin

117

would have been appreciated before the worst panic demands for treatment had set in. And had Koch not used his authority, as one of the most creative biologists of his age, to persuade others to accept his mystery medicine on trust, the scale of the disaster would not have been nearly so great. A lesser man would have been forced to unwrap his product for full inspection and would not have expected to hide behind his credentials.

There were few places Koch could look for support. In his own laboratory he had excluded from his colleagues the work which most needed the help of sceptical minds. Behring could be forgiven if he saw hypocrisy in the advice not to publish the work which was now revealed as far less premature than Koch's own. As for personal solace, there were few places for Koch to turn: he was poised unhappily in the middle of a breaking marriage, and his only daughter was now married off to a young doctor who had administered some of the first tuberculin injections. In the spring of 1891 he took a holiday in Egypt and it was from there that he wrote, as if half expecting rejection, to the girl who was far younger than his own daughter, Hedwig Freiberg. The letter he wrote at 47 showed his feelings of vulnerability and inadequacy, just as did those in his passionate youth: 'As long as you love me,' he told Hedwig, 'I cannot be beaten down by the vicissitudes of fate. Do not abandon me now for your love is my comfort and the star to which upward I gaze.' Hedwig did not abandon him. She became his wife. And, when looking for the motive which made Koch put himself in the terrible position of appearing to dupe the world, it is necessary to ask how much the founder of bacteriology was influenced by the need for cash from the sale of tuberculin licences to finance a breaking marriage and a mistress of 19. 'You couldn't blame Koch,' said one American, 'but what on earth did she see in him?'

When the furore and the optimism surrounding tuberculin was being fanned to its climax, and Koch was swept up into the uncertain role of public hero, the question of State assistance for tuberculosis research was raised in the Prussian Parliament. Luckily for Koch the subsequent disillusionment could not stop the momentum of support for an Institute to house his work, in spite of the fact that some, like Virchow, were ready to point out that the new establishment was being set up with what seemed to be indecent haste.

By the autumn of 1891 Koch had his Institute for Infectious Diseases. To help staff it he successfully recruited some of the young research workers who had already put themselves in the front rank of their subject, including Ehrlich, Kitasato, Wassermann and, inevitably if reluctantly, Emil Behring. The tragedy of a few months earlier, if it was not mentioned, was not forgotten. Koch continued his work on tuberculosis, but the centre of attention in the laboratory, and the work which now held out most promise for spectacular results, was that connected with Behring's diphtheria antitoxin.

Behring was extravert and aggressive. There is no doubt that he was accepted as the natural leader of the group working on the diphtheria problem, and there is no doubt too that he took up the role as his due. Koch soon found that Behring's laboratory had become virtually an autonomous institute within the Institute; there was little the Director could do about it. Koch felt that the place was becoming more like a farmyard than a laboratory as he watched larger and larger animals being assembled, for the diphtheria research.

Behring had to get enough blood serum containing his diphtheria antitoxin to begin trials on human beings. Starting with an immunised guinea-pig he had got enough blood from it to inoculate more guinea-pigs. From there he had progressed to rabbits, and from rabbits to sheep. By Christmas night, 1891, more than a year after publishing his work, he had enough antitoxin, and was sufficiently confident to take a needle and inject a small child lying in a Berlin hospital. This, the first of a series of experiments with human beings, was a qualified success. In a careful assessment of the therapy at the Berlin Charité Hospital – careful in the light of Koch's experience with antitoxin – the survival rate in patients was found to be high provided the treatment could be given early in the disease's progress.

Back in the Pasteur Institute, from which Behring had originally stolen the initiative in diphtheria work, Emile Roux was now using still larger animals. He was inoculating a horse with diphtheria toxin which had been weakened with iodine. Gradually he accustomed the horse to stronger doses until it was able to resist pure toxin. He then bled it from the jugular vein and was able to separate enough powerful antitoxin to begin treating children.

At the Enfants Malades Hospital from 1890 to 1893 there had been over 2000 deaths from diphtheria: a mortality rate of 51 per cent. In four months of 1894 Roux' work brought this down to 24 per cent. At the Trousseau Hospital in the same period, where the serum was not used, mortality was 60 per cent. This was success in the Pasteurian tradition. But in the publicity that followed, Roux was careful to see that Behring got his share of the credit. The French, by now very familiar with the unseemly rows over scientific ownership which the walls of Koch's Institute had not managed to contain, were cautiously avoiding contamination by the same unpleasantness. *Querelles d'Allemand* the French called the squabbles, much as the English prefer to dub a foreign origin on anything they find vaguely distasteful, such as Asian flu, Dutch courage or French leave.

Roux had scrupulously followed the French pure scientific tradition, well demonstrated in Pasteur, of refusing any opportunity of personal financial gain by exploiting his discovery. Behring was tied by no such tradition. In his experience, custom, such as it was, was to take the cash while the going was good: and Behring of all people needed no urging. Already, he had patented his diphtheria antitoxin.

In addition to Kitasato, without whom the serum might never have come into being, Paul Ehrlich was the other man who had helped Behring's work into a useful medical tool. This untidy but brilliant experimentalist had spent several years in and around Koch's laboratory. His habits and his method of work, relying almost entirely on test-tube experiments, used both to infuriate and amuse his co-workers. He wore detachable cuffs on which he made copious notes, smothering his waistcoat with the droppings from his thick Havanas as he agitatedly moved about the laboratory. His simplicity in his style of work helped him achieve some of his best results; his simplicity in his human relations was frequently childlike. Amiable and contented when engrossed in his work, he was capable of infantile outbursts, which added another upsetting ingredient to the charged emotional atmosphere of Koch's laboratory.

In Ehrlich's case personal financial gain was not a motivating force in any of his actions; he was too fascinated by the process of scientific discovery to think that there were delights to be bought from its profits. However, he was well attuned to the laboratory's custom of establishing scientific priority. There were many pieces of work to which he had made significant contributions of which he could feel proud and for which he could expect recognition. He had done pioneering work on the staining of bacteria and on immunity. In the diphtheria work he had used goats and horses to obtain antitoxin many times more powerful than Behring had succeeded in separating and, after a meticulous and long series of experiments, had discovered a method of measuring the amounts of antitoxin in blood, and calculating how much of it was needed to effect a safe cure in humans.

Ehrlich, then, could rightfully expect some sort of share, if rewards for contributions to the discovery of an effective diphtheria antitoxin

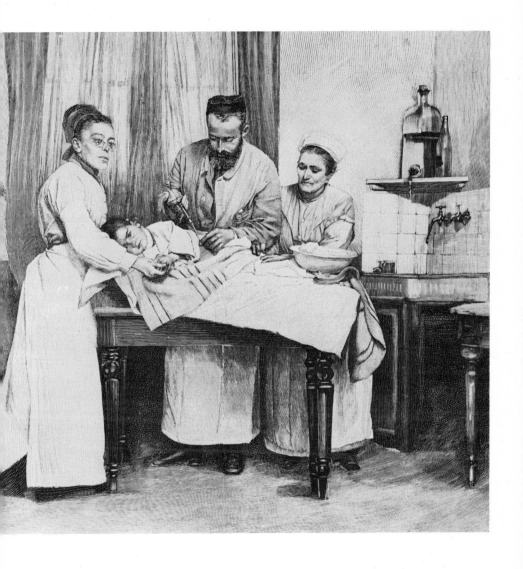

were being doled out. In the passage of antitoxin from laboratory bench to hospital therapy events moved with extraordinary speed and the early optimism was not misplaced. Commercialisation of the antitoxin began in 1892 when Behring was approached by representatives of the Lucius and Bruning dye works, at Höchst near Frankfurt. Behring agreed to have his product developed and marketed by the Höchst firm, though Ehrlich's rightful share of the profits needed to be settled. Behring had a solution, of sorts. By this time he had already cultivated useful contacts in the Ministry of Culture and he and the manufacturers were able to persuade Ehrlich to renounce his patent claim, in exchange for which Behring would engineer a salaried government post for Ehrlich, and one commensurate with his undoubted abilities, though being a government post, one in which he would be unable to accept royalties on patents. With naïve trust Ehrlich agreed. Behring, for whatever reason, was never able to use his charm with sufficient effect to fulfil his promise. Ehrlich had to suffer, not altogether silently, and watch a vastly profitable enterprise develop, from which he benefited by not so much as a pfennig.

Ehrlich hid his bitterness from no one. He took the matter to Prussian ministerial level and had no compunction in displaying Behring's inadequacies as a scientist. Intellectually Behring was Ehrlich's inferior, and the incident provoked Ehrlich into saying so in no uncertain terms. 'He wanted to be a Superman,' he wrote of Behring, 'but – thank God – he did not have the necessary super-brain.' The two never spoke to each other again.

Nevertheless, from what he had achieved so far with what Ehrlich considered limited scientific abilities, Behring reaped a very satisfactory harvest. The success of diphtheria antitoxin gave him both fame and fortune. For a time the fame overshadowed that of Koch. His ability to be influential at ministerial level served his own cause better than it had Ehrlich's, winning a professorship as well as ennoblement, with the right – which he exercised with pleasure – to call himself *von* Behring.

His gambling streak never deserted him, nor did his luck. On one occasion he swept the boards at Monte Carlo and he bought himself a villa on Capri with the profits. There he entertained lavishly, as he did at his home in Marburg: a castle overlooking the town, in which he built his own laboratory – and mausoleum. Nearby he established his own chemical factory: the Behring Works. And to help him share in the material success which, still only in early middle age, he so patently enjoyed, he married the daughter of the director of the Berlin Charité Hospital, who was then just twenty.

Ehrlich's view of Behring was, not unnaturally, coloured by events. Behring's style was personal, acquisitive and aggressive, but it was not vindictive. When old and crippled, leaning heavily on his cane, he was to be one of the scientists who walked behind Ehrlich's coffin to his grave where he said, 'If we have hurt you, forgive us'. In spite of Ehrlich's assessment, the contribution made by Behring to the signal triumph of conquering a terrible disease was great. And before he too died he was to make more contributions to the improvement of diphtheria immunisation techniques, and to work for many years attempting to find a solution to the tuberculosis problem. But even here there were disputes among former colleagues of the Koch camp. Still racing to be first, Behring applied for a patent for a tuberculosis immunising agent, only to see it disallowed in favour of application for a different agent made by Koch himself. Ehrlich took the trouble to write to the Prussian Minister of Education to make sure that it was realised at Governmental level that Behring's work was unoriginal.

In the event neither method worked and a satisfactory treatment for tuberculosis never came from a German laboratory, in spite of the years of creative talent, sheer hard work, tragedy and bitterness that had been associated with it. Yet again, the pendulum of success swung to France when Albert Calmette (who eventually became director of the Pasteur Institute), and Camille Guérin, after years of work together, successfully tried their BCG vaccine (bacille de Calmette et Guérin) on a baby delivered of a mother suffering from tuberculosis. The vaccine was developed using the microbe attenuation methods so brilliantly developed in Pasteur's laboratory. In 1921 the first successful trials began, having been delayed many months by the war. It was a war in which Calmette's wife was taken prisoner and held as a hostage, and in which Behring was again rewarded, this time by a decoration from a grateful German government which had used his tetanus antitoxin with sweeping success on its wounded soldiers.

Paul Ehrlich

CHAPTER NINE

No matter what views Paul Ehrlich held about the intellectual inadequacies of some of his colleagues, their views of him were united: behind the eccentric façade of 'Dr Fantasy', as they called him, there was an exceptionally able mind capable of working brilliant pieces of practical, creative chemistry. As a young research worker he had become fascinated by the preparation of dyes, and spent hours using them to stain animal tissues so that he could see them as they had never been seen. He wrote, for example, of his enchantment at being able to look at the fine twigs of the nerves of a piece of frog's tongue after it had been stained to a magnificent dark blue by the dye, methylene blue. His dyes became the centre of his life: his work and his hobby. Even when, in the evening, his wife played selections from popular operettas on the piano he, being tone-deaf, used the moment of relaxation to conjure up new thoughts about the use to which he could put his chemical compounds.

Ehrlich had been one of the first clinicians invited by Robert Koch to try out tuberculin on patients; together they had seen the severe limitations of the serum. Koch had quickly recognised the younger man's skilful techniques, and it was Ehrlich who perfected methods of staining the tubercle bacillus. During this period he contracted the tuberculosis that forced him into two years' convalescence in Egypt. On Ehrlich's return, Koch offered him a place in his Institute for Infectious Diseases, and there he began his work with Behring on diphtheria antitoxin.

Ehrlich's part in Behring's success was as Ehrlich himself assessed it to be: crucial. For a time it diverted him from his all-consuming love-affair with his chemical compounds but, in 1896, his abilities by now widely recognised, he was invited to direct the new Institute for the Investigation and Control of Sera in Berlin. At last, and without Behring's help, he had the control of his own Institute. As his own master, he could push his work along whatever channels his peculiar fancy took him.

Not that the new Institute was a model establishment with unlimited facilities: the laboratories were a converted bakery. For his techniques, however, Ehrlich needed little more than rows of empty shelves where he could store his bottles of chemical compounds; test tubes, water, a tap, a flame and some blotting paper were all the other equipment he required. It amused him to be known as 'the virtuoso of the test tube experiment'. His methods were highly refined, and he carried out these experiments with enormous care and with all the obsessive concentration required of a good chemist, being willing to spend hour after hour in repeating the same process. He knew the characteristics of the dyes in the bottles on his shelves as others might know the strengths and the weaknesses of human beings. And when he was depressed or in the middle of one of his feuds he would go and look at his bottles and say, 'These are my friends who will never leave me in the lurch.' Ehrlich attributed much of his success to being able to store away information about these compounds in the recesses of his mind and bring it out at the appropriate moment.

He had mixed views about the reasons for his creative triumphs and he was fond of quoting 'the four big Gs' as the essential constituents of any substantial scientific achievement: 'Geduld, Geschick, Geld und Glück,' (patience, ability, money and luck). All who worked with him saw the first of these two qualities in operation. As for money, he had no wish for it for its own sake, but he had seen at close quarters how important it was to Koch and Behring in furthering their careers and their laboratories' output. When the widow of a wealthy banker, Georg Speyer, financed a second institute for him to direct, to be built next to his existing one, any problems arising from want of simple equipment or staff were effectively removed.

As and when it suited him Ehrlich acknowledged the importance of luck in his discoveries. For example, he was fond of describing how once during his researches on tuberculosis he had tried unsuccessfully to stain red blood corpuscles. One day he had prepared his materials and put them down on top of a stove: typically, his laboratory was so cluttered with test tubes, bottles, journals and bundles of paper – he hated throwing anything away – that there was nowhere else he could see to put them. Next day, when he went into his laboratory he found that the woman cleaner had lit the stove. In a fury he retrieved his slides. But when he looked at them he found the most wonderful results; heating had brought about the staining process. But Ehrlich, even though he considered himself a lucky man, was ready to admit that the mere fact of a fortunate accident contributes little to scientific creativity; what gives it meaning is the store of knowledge from a mind capable of building on the fact; he, and he was never modest on the subject, had that store of knowledge. However, he had no suggestion to make as to how he ordered the knowledge in his mind, and the best he could do was to take refuge in the new psychology of Sigmund Freud, and suggest that he possessed a useful ability to store information at an unconscious level. Knowledgeable chemists who watched Ehrlich at work, however, were not deceived by the Chaplinesque performance – the detachable cuffs forever slipping over his wrists and the perpetual shower of cigar ash.

They confirmed that, in spite of the indescribable mess of his laboratory, luck had only a small role in his successes; he proceeded only after the most careful planning, and then with infinite care and persistence.

The laboratory atmosphere he created was, to say the least, singular. It was permeated by his cigar smoke, and by his voice ringing down the corridor, calling for a fresh box, or for mineral water with which to satisfy his apparently insatiable thirst. Aside from chemistry his intellectual stimuli were few, and these undemanding. Just as he drew inspiration from his wife's piano renderings of Viennese light opera in the home, so too in the laboratory he found he could concentrate better with the encouragement from the street of an organ-grinder, whose art he financed by the generous supply of coins. Those who worked with him looked on many of his characteristics as those which could be expected of a child. His naïve enthusiasm and his boundless energy inspired these same qualities in those who came to work with him. As one friend said, 'Ehrlich is a man whom one can love as a child is loved'. On the other hand, his childlike bad temper, his irrational rages and his conviction that all who were not friends were enemies figured strongly on the debit side of that same uncontrollable temperament. One of his habits was to scribble out his orders and his thoughts for the day onto small, differently coloured cards which he would have his assistant distribute to the different staff members in his laboratory. The cards were duplicated so that he could check that his orders had been carried out precisely. These cards, often nearly illegible, enshrined some of his most original thoughts applying chemistry to medicine. But delivered, as they were, as instructions to grown men, they were the cause of annoyance and, eventually, bitter resentment.

Ehrlich was unconcerned by the effects of his emotional outbursts on those he worked with. When his work on diphtheria antitoxin was over, he was able to return to all that really mattered to him: the use of chemistry to attack disease. He believed that chemical compounds could be used as real curatives, not, as they were then, as palliatives, acting on symptoms. He described his theory precisely when he gave his speech opening his new Institute, the Georg Speyer House, on 6 September 1906. That day he described the substances which would seek out and destroy the living microbes in the body as *magic bullets*. He spelled out the task of the new laboratory as being that of finding 'substances which have had their origin in the chemist's retort' to cure infectious diseases. The method would be called chemotherapy.

Looked at from today's vantage point Ehrlich's prediction that his Institute would give birth to a new way of treating disease is remarkable.   127

The original (and incorrect) formula for atoxyl. It supposed that the substance was a single side-chain attached to a benzene ring

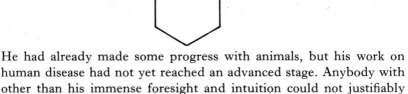

$$\text{—NH —As O}_3 \text{H}_2$$

He had already made some progress with animals, but his work on human disease had not yet reached an advanced stage. Anybody with other than his immense foresight and intuition could not justifiably have made such a sweeping prophecy.

A turning-point in Ehrlich's work came when it was found possible to infect rats and mice with trypanosome microbes. With what can certainly be described as intuition, but with what was in fact the result of a phenomenal knowledge of the behaviour of chemical compounds, Ehrlich had plucked a substance from his shelf and had had one of his assistants use it on an infected mouse. A derivative of this substance, a dye he called trypan red, had killed the trypanosome and cured the mouse. Burns' stanza beginning 'The best laid schemes . . .' might well be brought to mind by Ehrlich's determination to extrapolate from mouse to man, but he was aware, more than most, of the formidable problems involved. There was immediate hope that trypan red might be of use in the treatment of the trypanosome which causes human sleeping sickness; but, when the dye was sent abroad for doctors working in the tropics to try on their patients, results were too variable for it to be used in treatment.

However, whatever trypan red did not do, it also did not kill the patient. The results of his early work gave Ehrlich grounds for optimism and intense activity. He called the hypothesis on which he was basing his research his *side-chain theory*. This was the controversial theory he had first introduced in his prodigious and painstaking work on immunity. Now applying it to microbial infections, Ehrlich believed that the chemical compounds he was using would only effect a cure if there was a particular relationship between the substance and the microbe in question. The one must fit the other as a key does a lock. He believed that, attached to the molecules of the microbial cell, there were groups of atoms which stuck out like arms, and that the problem of dealing with the microbe could be solved by finding a chemical compound of a suitable shape to lock onto this arm. And to be of any value in therapy, the substance must not harm anything else in the human body.

Ehrlich already knew that another substance called atoxyl, an arsenic derivative, had shown some promising results in trials on sleeping sickness. He also believed that the formula of atoxyl, which had been worked out and accepted several years earlier, was wrong (above left).

But, again through his great store of knowledge – being familiar with the sort of chemical reactions atoxyl could, and could not undergo – Ehrlich guessed that the nitrogen atom was misplaced in this formula, and that the true formula had not one, but two, side-chains (above right).

128

$$H_2N - \text{(benzene ring)} - AsO_3H_2$$

The original formula led chemists to believe that atoxyl could easily be decomposed and that its structure was not well suited to having new groups of atoms attached. Ehrlich's formula, however, suggested it was a stable substance which could be manipulated, and that groups of atoms could be stuck on to it to give new side-chain derivatives. If he was correct, it should be possible to arrive at derivatives which, in the treatment of sleeping sickness and other diseases, might be more effective than the original atoxyl.

Ehrlich responded to his own thought in the only fashion he knew. He scribbled instructions to his chemists on one of his coloured cards. The information it carried – that, for no particular reason, the formula of a chemical compound should be assumed to be something which they and establishment chemistry knew it not to be – reeked of scientific dogma.

The incident soon took on the proportions of a first-class row, and Ehrlich found himself facing three angry senior chemists of his laboratory, Drs von Braun, Schmitz and Bertheim, who were sufficiently incensed to challenge the untidy man's dogma. But in a confrontation of this kind Ehrlich was unshiftable. By now in one of his fiercest moods he told them without any ado that, in his laboratory, they would do as they were told. 'You,' he shouted at the three well-qualified men in front of him, 'cannot judge whether this is right or wrong,' and walked out on them. Von Braun and Schmitz resigned on the spot. Alfred Bertheim, counting discretion and uncertainty as the better part of valour, decided to wait and see. It was a decision he never regretted. He found that Ehrlich was utterly correct. Stable derivations of atoxyl could be produced in the laboratory. The textbook formula was wrong. Just as Toscanini, more than a hundred years after the composition of a piece by Beethoven, as a result of his deep intuitive knowledge of the composer, could spot an acceptable and accepted copyist's error in a single note; so by similar mental process could Ehrlich spot a flaw in the beauty of a chemical formula.

Bertheim's star was to rise to success with that of the Master. His luck, however, did not have the consistency of Ehrlich's. It fouled in an incident which would have been uproariously funny but for its pathetic outcome. When the Great War broke out, Bertheim joined the army. It was to be a short war for the poor chemist: the fish out of water. On his first day with his unit, sporting his dress uniform, he began to descend some stairs. Sadly, his spurs tangled with the carpet and he fell to his death with a broken crown.

Now that atoxyl was properly identified Ehrlich showed how useful the money – which he had always seen as an essential condition of success – could be. In his new Institute, securely financed by the accommodating Frau Speyer, there were ample staff and resources for Ehrlich to embark on a long line of experiments with derivatives of atoxyl. The number of compounds they made went up by tens and hundreds and eventually approached a thousand. Each compound required meticulous preparation and equally meticulous testing on animals. With the 418th substance came optimistic signs; it was effective on spirochetes, the spirally twisted microbes which caused syphilis.

One of the problems of trying to find an effective cure for syphilis was that it had been found impossible to infect a laboratory animal with the disease. But in 1909 a young Japanese scientist, Sahachiro Hata, came with a recommendation from Ehrlich's old colleague, Kitasato, to work in Ehrlich's laboratory. He was reputed to have discovered a method of giving syphilitic infections to rabbits. Outside microbiology the ability seems scarcely desirable; to the initiated it was a quite exceptional achievement.

In these years, at the beginning of the century, the disease was unmentionable in most circles outside laboratories. There the thing could be looked on as a medical curiosity, rather than the ruin (which it was) of so many useful human lives. In cold statistical economic terms, its effects were startling enough. In every country in the West each year the number of man-years lost as a result of insanity, heart disease, paralysis and blindness caused by syphilis reached hundreds of thousands. In Germany six per cent of all deaths were due to syphilis; in France, ten per cent. For the individual its effects were terrible. When Ehrlich's chemists were looking for some clues to a treatment one doctor talked of wards full of soldiers, their faces rotting away from the tertiary stage of the disease. Only one in five ever recovered from it. It was not simply contagious: it was congenital. Moral attitudes to it were peculiarly ambivalent. When children inherited it, some thought it right that the sins of the fathers should be visited on their sons; they considered that a cure would be open encouragement to immorality.

When Hata arrived, bringing with him his special knowledge of the delicate corkscrew-shaped microbe, Ehrlich handed over two of the compounds from the long list he, Bertheim and others had over the years produced from atoxyl. They were 418 and 606. There were high hopes for 418, because of its action on some spirochetal infections, but 606 had been tried by one of Ehrlich's assistants and found to be ineffective. It was therefore almost a disagreeable surprise to Ehrlich when Hata turned up in his office one day and, bowing low as usual, announced that

606 was effective on the spirochete of syphilis that infected laboratory rabbits and chickens. Ehrlich was not the first, nor will he be the last director of research to be faced with entirely conflicting experimental results from two assistants, and not know which to mistrust most. But he allowed his intuitive processes to act as they normally did. The stream of abuse he let out over the bowed head of Hata was aimed at the absent and incompetent assistant. Hata's results, under repeated testing, stood firm.

After Ehrlich was satisfied with his laboratory trials two of his colleagues' assistants volunteered to submit themselves to doses of 606 to test its safety. This was an anxious time, since blindness was known to be one of the side-effects of some arsenical preparations. The young men survived unharmed. However, the first patient to be treated with salvarsan (as 606 was to be called), was suffering, not from syphilis, but from a disease caused by a different organism, relapsing fever. A St Petersburg doctor, Julius Iverson, had shown that atoxyl was effective in the treatment of this disease, and Ehrlich therefore sent him samples of 606 to try. Of 55 patients treated by Iverson during the city's relapsing fever epidemic of 1909, 51 recovered totally after one injection. Equally dramatic effects occurred with cases of syphilis, some of the hideous, disfiguring ulcers disappearing within hours of the first treatment.

Ehrlich, however, was cautious. He had no intention of making a premature announcement in the light of Koch's experiences with tuberculin. Indeed, many relapses did occur in syphilis treatment when overcautious doctors administered insufficient doses. Loss of hearing was one frequently reported worrying side-effect. But by examining in detail the work of all doctors who were using salvarsan, Ehrlich was able to reassure himself that the sometimes permanent injuries were the consequence of the misuse of salvarsan, and not one of its unavoidable side-effects. By 1910 he was sufficiently confident to announce his discovery.

Recognising that he had achieved what he forecast could be achieved from chemistry, he ordered the whole of his chemical laboratory to do nothing else but prepare more of the yellow, crystalline powder of pure 606 until the time that the Höchst chemical works could begin large-scale production. Ehrlich himself kept a record of the doses handed out each day on the back of the cupboard door in his cluttered office. They ran into many thousands and Ehrlich was often to be found on his hands and knees as the list lengthened towards the carpet. He allowed no doctor to be supplied with the compound unless he undertook to keep him, Ehrlich, informed of its clinical results.

In spite of all his care there were more relapses, and some unpleasant side-effects which could not be prevented. Some of the problems arose from the way in which the yellow powder had to be administered. It had to be dissolved in a large amount of water literally at the bedside and then injected into a vein: a tricky procedure even when performed by a skilled doctor. If much of the solution escaped into the tissues around the vein it could kill off the living cells; the result might mean that the patient's limb would have to be amputated. But by 1912 Ehrlich had introduced his 914th derivative of atoxyl, *neosalvarsan,* a more soluble substance and a more reliable cure.

Ever since he had announced the discovery of salvarsan two years earlier, Ehrlich's life had taken on a different aspect. He had become famous. Salvarsan was ideal newspaper copy, and the mass-circulation papers were now enjoying their boom years. The new discovery and the disease it cured had almost every journalistic ingredient to make it qualify automatically for reportage under banner headlines; with sex, suffering, medical triumph and more than a loose connection with immorality, Ehrlich's work could not fail to have regular and frequent coverage.

However, once the swell from the wave of public reaction to the first announcement had died down a little, the tragedies of chemotherapy now began to qualify for a substantial number of front-page column inches. Some of the early disasters associated with salvarsan – the amputated limbs, the permanent deafness – were given full publicity, and the clinical trials of the drug came in for harsh criticism. Ehrlich took every piece of criticism personally and seriously, and attempted to deal with it as best he could. But by this time his health was not good; the exposure to a public which was not always either rational or even modestly well informed about his work took its toll on both his temper and his energy.

The longest period during which he had to put himself on public display in order to defend and explain his work was also the most bizarre. It began when Karl Wassmann, the eccentric publisher of a Frankfurt paper, *Die Wahrheit* (The Truth), used his pages for a series of attacks on salvarsan and on the administrators of the Frankfurt Municipal Hospital. Wassmann claimed that not only were the local consultants using a dangerous drug, they were making entrepreneurial profit out of it. He also made the startling accusation that the city's prostitutes were being dragged into the hospital, being forcefully treated with salvarsan and being used as human guinea-pigs in clinical trials.

Eventually, in May 1914, the city authorities took libel action against

Wassmann, and Ehrlich was called to the Frankfurt Court of Justice as an expert witness. The trial was long, unusual and, for Ehrlich, exhausting. During the hearing Wassmann and some of the prostitutes made a habit of interrupting the evidence from their seats in the body of the court. Unconventional as their methods were, there was no doubt that there were some worrying aspects to the case. There was irrefutable evidence that salvarsan treatment had been made compulsory for syphilitic prostitutes and forcibly given to some of them, and that over the years there had been four deaths in the prostitutes' ward; however, the causes of the deaths were in dispute. Giving evidence, Ehrlich had to admit that during the early months of its trials the unskilled or careless use of salvarsan had led to disasters. In the end, however, no evidence was produced to show that any doctor had made money from the promotion of salvarsan, and Wassmann was sentenced to a year's imprisonment. But although the action left a considerable number of questions relating to the desirability of forcible treatment of patients unanswered, it left the reputations of Ehrlich and his salvarsan intact. However, being made to defend his work in this way – in public, and in a bizarre setting – threw Ehrlich into a deep depression. It was a depression which began to affect his health, and from which he never properly recovered.

In truth, there was no rational reason for this great bout of introspection. He had been as good as his word. He had devised and planned a research programme which, after thousands of laboratory bench experiments and tests on animals, had been completed within six years. He had designed a chemical which could search out and destroy a microbe within the human body. His first success in this field had been to provide a wonderful cure for a hideous disease. In what to the casual observer might seem a completely disorganised and messy fashion he had devised techniques which would give a new orientation to medical research and to twentieth-century medicine. Ultimately the lives saved and the economic and social effects of this man's work are incalculable.

Ehrlich himself, however, was not satisfied with what he had done. His dream was to produce a substance which, with one big dose, could effectively destroy all the microbes of a disease which had invaded the body. Until days before his death, he was still looking for his solution, still working on salvarsan and still the childlike, irascible old man hunting out his microbes with the unswerving fierceness he used on any human whose opinion differed from his own.

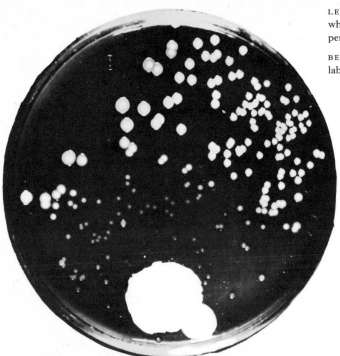

LEFT: The original culture plate on which the observation of the action of penicillin was made

BELOW: Sir Alexander Fleming in his laboratory in 1954

CHAPTER TEN

Paul Ehrlich died in August 1915 after a severe stroke. In his last months he was often deeply depressed: depressed about the work he knew he would never be capable of doing, and depressed about the war in which scientists would be heavily involved. Perhaps more clearly than most, Ehrlich realised that this war would be the first of the technological wars: one in which the character of the conflict would be governed by applied science. Salvarsan itself was widely used against syphilis in the Great War; in Britain, where Ehrlich's patent no longer had to be honoured, it was manufactured and used with great success under the name arsphenamine.

It was the first synthetic drug to be employed on a wide scale. It was the triumphant product of planned research and it heralded the beginning of a new era in medicine for which Ehrlich was entirely responsible. It also marked the end of an era; the likelihood is that history will never again see another as astounding and influential in the application of science to medicine. Fifty years earlier the germ theory had been a piece of guesswork: an uncertain tool in the hands of unknown chemists, biologists and doctors. Now it was a rock of knowledge. What could be built on it would amount to far more than a simple understanding of the cause of a disease; for understanding would lead to control, and control to a vast economic and social effect on society.

Ehrlich himself named the new branch of medicine that sprang from his work: he called it chemotherapy. The hopes that followed the success of salvarsan were of the manufacture of a medicine which, in a syringe, in a spoonful, or in a pill could be specifically designed to deal with whatever particular microbe caused an illness. However, the days of the pill society were some way away. For twenty-five years following the synthesis of salvarsan, nothing to match its spectacular action came from any of the dye industries that were turning themselves into drughouses in the hope of continuing where Ehrlich left off. A few synthetics were produced which improved on quinine as an anti-malarial drug, and there were some successes in the treatment of urinary infections. Then, in the early 1930s a German chemist employed by the I. G. Farbenindustrie, Gerhard Domagk, began to look at the effect of a number of dyes on mice infected with a deadly streptococcal infection. Domagk's first reports of his work with a new red dye showed it was effective in a very limited number of trials with mice. However, he had sufficient confidence in his work eventually to try this substance – prontosil was its trade name – on his own daughter who was desperately ill as a result of a streptococcal infection after she pricked herself with a knitting

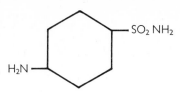

needle. Everything else tried by Domagk, a doctor by training, had failed. Prontosil produced a dramatic recovery in the girl. It was the first *sulpha drug* : the first of a string of compounds whose influence for good was going to be enormous.

By this time the Pasteur Institute had become interested in chemotherapy: there was now something of an inevitability about the swing of the pendulum, back and forth from Germany to France. In Paris a husband and wife working together, the Tréfouëls, noticed that prontosil killed streptococci only in animals or human bodies and not outside the body under experimental conditions in a test tube or on a slide. They therefore reasoned that it was possible that the body itself converts prontosil into a substance which acts on the microbes. Soon they were able to break down the prontosil into a much simpler molecule called sulphanilamide which they identified as the active constituent of the drug. It became the first of a series of compounds to which the phrase 'wonder drug' could be attached without much exaggeration. It and its derivatives were found to be capable of dispatching the infective agents of many diseases with an effectiveness which only sufferers who were once without hope, and who were now given a future, could truly appreciate (above).

Ironically for the German dyestuffs firm in whose laboratories prontosil was discovered and who were hoping for many years of cash return from their monopoly, sulphanilamide was not patentable. By introducing new groups of atoms into its side-chain, as Ehrlich had done with atoxyl, it was possible for anybody who was clever enough to arrive at a string of compounds, each with slightly different, but powerful action on bacterial diseases. A new race now got under way to manipulate the side-chain in order to make patentable products each with specific pharmaceutical effects.

Still one of the most severe and most cruel of streptococcal infections was puerperal fever. In spite of all the measures which had been put into operation as a result of an understanding of the germ theory, its spread in labour wards was merely curtailed, and it was far from being eliminated. In Queen Charlotte's Maternity Hospital in West London in the four years following 1930 the average mortality rate in women infected with childbirth fever was 22 per cent. The numbers involved were far less heart-rending than in the days when Semmelweis was operating at the Vienna General and Tarnier at the Paris Maternité, but they were still disastrously high.

In 1930 not much short of a hundred years after Semmelweis had

battled so ineffectively to put across the idea that might have saved untold numbers of lives, Leonard Colebrook and his co-workers deduced, and demonstrated practically, the principles on which Semmelweis's assumptions were made. Colebrook showed that the streptococci which cause the disease do not become dangerous unless they can grow on bruised or torn human tissue. In other words, a prime factor in preventing puerperal fever is to ensure the utmost gentleness in obstetric operations. It was also shown at this time that the streptococci can be harboured in the nose and throats of the staff in the delivery room. Until then the obstetrician and his assistants did not wear either masks or sterile gloves at a childbirth. Only now was it possible to see the true reasons for the appalling deathrates in the teaching ward of the Vienna General and over which Semmelweis had so grieved.

However, the actual treatment of a woman suffering from puerperal fever was still utterly primitive. In Vienna she would be given an intravenous injection of alcohol: in London a stiff brandy. The effect in both cases was to make the patient drunk; that apart, the treatment was ineffective.

Improved surgical procedures reduced the incidence of puerperal fever during the early years of the twentieth century. But even in 1935 deaths from the disease were still very significant. In that year, while he was working at Queen Charlotte's, Colebrook was given a supply of prontosil and, when an unexpected epidemic of puerperal fever broke out, he began to use it on his patients. In the early stages of the epidemic ten women had died: a mortality rate among those infected of 26 per cent. Colebrook, in his first trial, cut the mortality rate to 8 per cent. It had taken many years of the germ theory in application to bring this killer of the labour ward down to something less than frightening proportions.

By 1938 another sulpha drug, sulphapyridine, had been tried out on a Norfolk labourer with severe pneumonia; although in an apparently dangerous condition, he recovered. He was the first success of 'M and B 693', a drug manufactured by the British firm, May and Baker. The diseases against which it had spectacular success included pneumonia, meningitis, gonorrhoea and staphylococcal infections. And when it was administered to Winston Churchill during the Second World War and he found it not to be wanting, he took up the metaphor of the microbe hunters in well-rounded, pugnacious phrases. 'This admirable M and B,' he said, 'from which I did not suffer any inconvenience, was used at the earliest moment, and, after a week's fever, the intruders were repulsed.'

Bacteria have collapsed before the assault of chemotherapeutic agents,   137

$$H_2N - \langle\ \rangle - CO_2H$$

as Churchill or Metchnikoff would have it, as invading enemy battalions have fallen to the home forces. By what mechanism this occurs, however, is still not certain. Ehrlich's theory that the agent fitted something attached to the microbial cell in a similar way that a lock fits a key was soon found to be inadequate. The kind of mechanism that is involved is likely to be connected with the metabolic processes of bacteria. One of the key substances in this process is folic acid, and one of the compounds which is used in its synthesis is another acid, para-aminobenzoic acid.

Now there is an easily recognised similarity in shape and constitution between para-aminobenzoic acid and sulphanilamide (compare figure above with figure on page 136). If, therefore, bacteria are grown in the presence of sulphanilamide, rather than take up para-aminobenzoic acid to form folic acid, they will take up sulphanilamide. The result will be that the chemical compound they form will not function in the same way as folic acid and the bacteria, denied this substance essential to their metabolism, will fail to multiply. In other cases other mechanisms are involved but, even in these, it is likely that the drug concerned interferes with the metabolic process essential for the survival of the bacterium.

Had the methods of scientific research and development at that time been the same as those in operation during most of the hundred years or so before their discovery the sulpha drugs might have had a long life as the most potent and widely used of materials in man's fight against the microbes that, seemingly of right, had invaded his body. But new techniques which improved the artistry of the chemist, and broadened the scale on which he could operate, meant that in the end they had only a short reign. Even before this period, experiments which would bring into being new substances to dominate the chemotherapeutic scene, and which would be of an entirely different origin from any yet used in medicine, were under way. In what is now probably one of the most frequently described incidents of scientific history, a Scottish bacteriologist working in a London teaching hospital, Alexander Fleming, had already watched the sweeping effects on microbes of a substance produced by a living organism.

The words 'pure chance' have probably also been applied more often to what happened that day in 1928 in Fleming's laboratory in St Mary's Hospital than to any other important scientific event. However, the actual mechanism of that chance will probably be a bone much chewed over by scientific historians for as long as the history of the subject is written. Fleming had already carried out some interesting work in which

Dec 11. 26

Inhibition by moulds.

Moulds planted on broth in flasks nov 30.   Room temp. (cold weather)
Equal parts ~~with~~ mould broth and boiling agar mixed and filled
into holes in an agar plate.   After solid. surface flooded with
blood agar containing haemolytic streptococci.
After 18 hours.

mould broth imbedded.   Inhibition of growth

1. Inhibitor   Complete for 5 mm. round

2. C Jamia Viridescens   Partial only over imbedded broth

3   Botryti· cinereum   nil .

4   Aspergillus fumigatus   nil .

5   Penicillium   complete for 7 mm around.

6   Sporotrichium   nil .

When inhibition of streptococcal growth then blood corpuscles
preserved. Others laked.

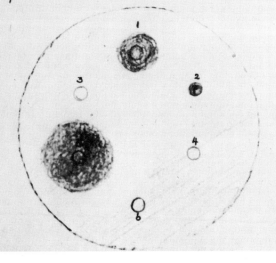

Feb 16 26

Shows
(1) Inhibition of staph by mould filtrate
(2) Symbiosis B. Influenzae & Staph.
(3) Increase of colouration of staph by proximity of one mould & not of another
(4) Zone of inhibition of staph & Pfeiffer close to margin of area where mould applied.
(5) Existence of white & yellow staph together.

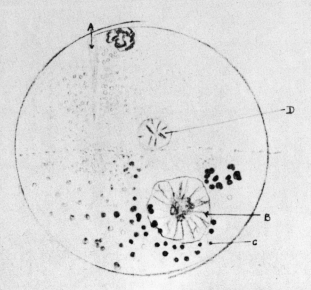

Blood agar plate   Craddock's nasal mucus equally spread & shew above.
6 drops mould filtrate applied to upper half (above red dotted line)
In 24 hrs growth of yellow white staph. and B. Influenza in lower half (comparatively few B. Influenza among staph but many close to margin of mould filtrate where staph beginning to be inhibited.   Many microscopic colonies of Pfeiffer on upper half (medium not good so colonies not seen with naked eye.)
After 2 days stroke of staph made in upper half (A)   Incubated 24 hrs and mass of B. Influenza colonies came up round staph large near staph and tailing off.
Staph tails off towards margin of mould filtrate and a zone free from colonies visible to naked eye. along margin of mould filtrate
After fortnight on blood agar large mould grows in lower half and round it for 2 cm. staph have a very pronounced increase of orange colour. (C)   Another mould (D) has no such effect

he had shown that an enzyme present in tears, and in mucous from the nose, lysozyme, can kill some kinds of bacteria. He had a habit of cluttering his laboratory with culture dishes long after the time that other workers might have thrown them away. On the day in question he noticed that a culture of staphylococci, the bacteria which form pus, had become contaminated with a mould. Again, a worker interested in effects of a nature quite different from those that intrigued Fleming would have thrown the contents of the mouldy dish down the sink.

Fleming was notorious for keeping closed the windows of his laboratory, which looked out onto noisy Praed Street. Later he claimed that the mould spore had blown in from the street, though the likelihood is that it came up the staircase. Fleming did not know that, on the floor below, a young Irishman was working with strains of penicillium mould.

Fleming looked at his plate and saw that there were patches where the staphylococci had apparently been destroyed by some substance produced by the mould. The mould itself was identified by the Irishman, C. J. La Touche, as *penicillium rubrum,* but it was later shown to be *penicillium notatum.* However, like the customs of folk-history which predated Jenner's vaccination against smallpox, the mould's properties were not being revealed for the first time. In many parts of Europe country people would keep old loaves on the rafters alongside hams and bacons. They knew that the mould which formed on the bread's surface, if mixed to a paste with water, was somtimes very useful in preventing infections when bandaged over an open wound.

Fleming's training and experience was such that he might call himself a fully paid-up member of establishment science. As a result of what he saw in his culture dish he followed establishmental procedures and lived to thank heaven for the fact. He published his results. At the time nobody outside his own laboratory took much notice of his observations. Fleming did not have the chemist's skills with which to separate the substance that was so effective in dealing with microbes, nor did he demonstrate its chemotherapeutic effects. Not until ten years later, unknown to Fleming, did Howard Florey and Ernst Chain begin to isolate the anti-bacterial substance. This was penicillin. Before the end of the Second World War it was being produced in America by culture techniques on a vast scale, and being used with wonderful success against pneumonia, gonorrhea, scarlet fever and meningitis, and was replacing sulphonamides in the treatment of puerperal fever, and neosalvarsan in the treatment of syphilis. It was available in large quantities for use in the hospital lines after the invasion of Europe began. Without it the summer of 1944 would have seen rows of men stretched out suffering from wound infections, many of them dying, as they had in France in the battles of 1914-18.

This drug was the first of the antibiotics. As such its significance can scarcely be overrated. It was now plain that other antibiotics might exist. Just as the hunt for the microbe had revolutionised the understanding of the nature of disease, the hunt for antibiotics was to revolutionise treatment. By the early 1940s the techniques to identify and separate these organic products of living cells were well advanced. In chronicling the link between the work of the bacteriologists of the nineteenth century and their counterparts in the mid-twentieth the achievements of the latter will here receive perhaps superficial acknowledgment. The effects of their work on medicine and society, however, have been anything but superficial.

In 1944 streptomycin was isolated from a soil mould by Selman Waksman working at Rutgers University in the United States. Its greatest success was against tuberculosis. It was now more than half a century since Robert Koch had carried out his great work isolating the bacillus, since he had raised so many false hopes with tuberculin, and since he and Behring had disputed the patent rights of a so-called treatment for the disease. At last an effective cure for the romantic killer disease had arrived. The patent of streptomycin – an extremely valuable one – was placed in the name of Rutgers University.

Soil turned out to be an extraordinarily useful place to search for antibiotics. It provides an amazingly clean environment. Very few bacteria can survive in it, and it clearly contains many anti-microbial substances. A soil sample from Venezuela led to the discovery of chloramphenicol ('chloromycetin') in 1947. This was the first 'broad spectrum' antibiotic, useful against a wide range of bacterial infections. It was the first antibiotic to be synthesised: that is, to be tailor-made in the laboratory, as Paul Ehrlich had prepared his first chemotherapeutic agents. Chloramphenicol was found to be very effective against typhoid fever and many other infectious diseases. So too are the tetracycline group of antibiotics – aureomycin, terramycin, and others.

The list of successful agents which have been turned out in bacteriological and chemical laboratories to hit and kill specific microbes is, at this stage in a summary of their history, becoming a catalogue. To pretend that the battle is won would of course be nonsense. Tuberculosis, for example, is still rampaging through parts of Africa. But what has happened since the birth of the germ theory, when man entered serious conflict with infectious diseases, is that the state of ignorance of the

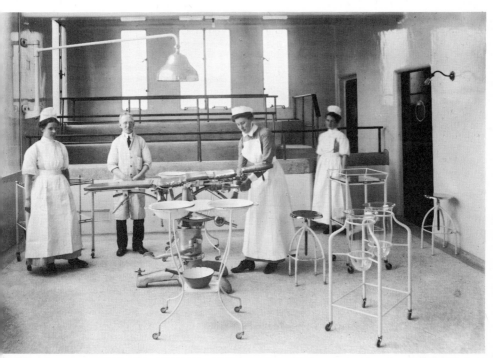

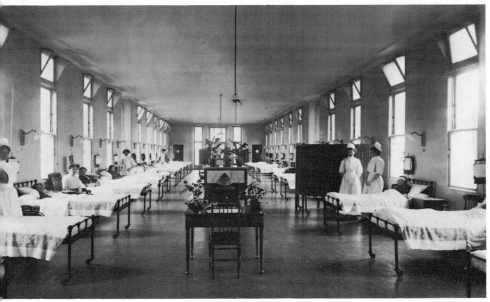

weapons he should use, and of why he should use them, no longer exists. In most cases, he has a good chance of victory, and in some he has parity. The metaphors of modern warfare – strategy, tactics, overkill, target elimination, etc. – can be drawn on at will to describe the techniques and consequences of modern drug theory. And just as the side-effects of warfare can be social and political problems on a vast scale, so too can be the unexpected results of the application of the products of the germ theory. But before looking at these – and they can scarcely be avoided – there are still problems to be solved that deserve mention.

CHAPTER ELEVEN

During his last series of experiments on rabies, Pasteur, in one of his extraordinary flights of intuition, based a theory, and the experiments that confirmed it, on the assumption that the cause of the disease was a microbe he could not see, or even verify was present. He decided to try to culture this micro-organism in nerve tissue under aseptic conditions. The method he wished to use was to involve inoculating rabies into a dog, the site of the inoculation being the tissues of membrane surrounding the brain.

Even to Pasteur, who had had to withstand bitter attacks on his work by anti-vivisectionists, and who, incidentally, was not particularly fond of dogs, the whole idea of an experiment on the brain of a living animal was repulsive. Roux was the brilliant, and unemotional, young man who actually performed this piece of work. When the dog had recovered from the experiment, he took it to Pasteur who 'lavished upon the dog the kindest words'. Two weeks later, as planned, the animal developed rabies, and Pasteur showed that it was possible to cultivate a virus – that was what it was – in the body of an animal.

For years nobody actually saw a virus. In 1897 a Dutch botanist, Martinus Beijerinck, pressed out the juice from tobacco leaves infected with 'tobacco mosaic disease', which gives the plant a mottled appearance. When he passed the juice through a porcelain filter he found it was still capable of infecting plants, but he could not culture any bacteria from it. Beijerinck came to a sweeping conclusion from his limited observations; he believed there was an agent of disease dissolved in the liquid, and he first gave it its name, calling it a 'filterable virus'. He had done little more than make a guess, but it was more or less correct.

In the same year two of Robert Koch's pupils, Friedrich Löffler and Paul Frosch, were investigating foot-and-mouth disease; they too were able to filter off a liquid which was capable of passing the disease from one cow to another. However, they suspected that the agent responsible was not, as Beijerinck had believed, something that had dissolved itself in the liquid. They believed it to be an organism much smaller than a bacterium: so small in fact, that it had passed through the filter.

Within thirty years several dozen human diseases, including smallpox, chicken pox, measles, polio and yellow fever, were known to be caused by viruses. Still, however, there was a wide ignorance as to what a virus actually is. The main problem preventing the creature – or thing – being brought out from behind its veil of mystery was that there was no way as yet of studying it. Pasteur had shown how a virus could survive and multiply in a living tissue; the difficulty lay in getting it to stay alive outside the body. By 1913 a technique had been developed of keeping cowpox virus alive on pieces of cornea from rabbits and guinea-pigs. But animal experiments take time and are expensive. A reliable, inexpensive method was required which would make it possible to study viruses on a large scale. Eventually, in the late 1920s, a wonderfully simple aid was found: the hen's egg. It was discovered that chick embryos, protected by their shell, and built to deal with the bacteria which otherwise interfered with the process of investigation, could be used as self-contained units to grow many different viruses. This product of the poultry farm would eventually make the manufacture of virus vaccines a practicable possibility.

But at this stage the nature of the virus was totally hidden. It was still inviolable. The wonder-drugs – both the first successes of chemotherapy and the first antibiotics – left it untouched. There was considerable debate about whether the virus was alive or dead. Then, in 1935, Wendell Stanley at the University of California, with enormous technical skill, actually succeeded in crystallising the tobacco mosaic virus. He found that, when he redissolved his crystals in water, they were just as capable of passing on the disease. It was a remarkable piece of work.

It now looked as though the virus was not a microbe at all, but simply a protein: an inanimate complicated molecule. But here was a dilemma. Whatever the thing was, it could grow and reproduce with all the dynamic qualities of life. Yet it could exist in the solid, well-formed stable shape of a crystal, as unmistakably unalive as a piece of salt. The virus, it seemed, hung somewhere between life and non-life.

But it suddenly achieved new character and new importance when two British biochemists, Frederick Bawden and Norman Pirie, showed that the tobacco mosiac variety was made up of 6 per cent ribonucleic acid, RNA. Later, other viruses were shown to contain small but altogether significant amounts of either RNA, or DNA – deoxyribonucleic acid. What was important was that this is the material of the genes, the substance of life. Any difference of opinion of whether the virus was alive or not-life was not a question of mere semantics, but one of what is or what is not life? Now that some of the molecules essential to life had been discovered in the virus it was clear that to study the fundamental nature of the mystery was to study the fundamental nature of life.

Knowledge of a big issue is, therefore, embodied in this small and, for man, vicious object. Viruses range in size from the large cowpox virus, which, like bacteria, can be seen under the optical microscope, to the minute polio virus, a roughly spherical object, a million of which

LOW LEFT: Tobacco mosaic virus

LOW RIGHT: Polio virus

TTOM: Cowpox (the dark areas) growing on egg membrane

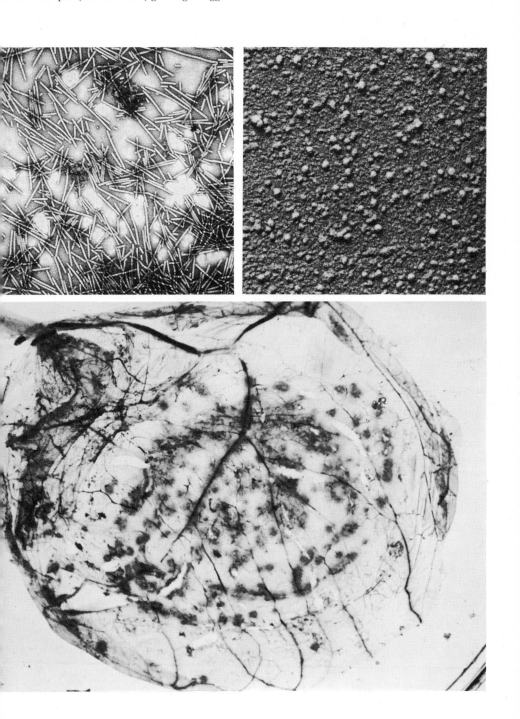

could comfortably fit side-by-side in an inch. The smallness of so many viruses is one of the characteristics which, from the start, has limited knowledge of the species. Not until the late 1930s was it possible actually to see some viruses; only then were the techniques of physics so refined that physicists were capable of getting sufficiently huge magnifications with an electron microscope to be able to see the shapes of the thing biologists were dealing with.

Viruses exist and stimulate the host cells they live on to reproduce them parasitically. Some feed on the cells of the body, and some even on bacteria; these last types are called *bacteriophages*, or bacteria-eaters. The problem for medicine is to strike at the virus without striking at the living cell. The major strategy of dealing with human viral diseases, one which Pasteur had adopted from the start, has been that of prevention rather than cure. Seeing where his work could be most effective, Pasteur concentrated the work of the later years of his life on vaccination.

What clearly had surprised Pasteur when dealing with rabies, as much as it had surprised Lady Mary Wortley Montagu and Edward Jenner when dealing with smallpox and cowpox, is the amazing ability of the body to withstand viral attacks. The virus can have a calamitous effect on individuals and on whole communities. And yet, remarkably, considering its pernicious nature, large numbers of people in closely packed communities, many of whom had not been medicated, nor taken any really useful protective measures, survived the smallpox epidemics of the early nineteen hundreds, and the influenza pandemic of 1919 unscathed. It is obvious that, under certain favourable circumstances, the human body possesses an ability to deal with the virus by its own unaided mechanisms. When it is vaccinated the body's ability to resist infection is even further extended in a most wonderful fashion. What is this mechanism?

The idea put forward by Metchnikoff – with his splendid vision of ranks of heroic white-uniformed phagocytes surrounding the wicked black bacterial hordes – was a mite too simple to be true. It provided a wonderful allegory, and the picture it invokes of 'us' and 'them' with 'us' permanently trying to beat 'them' off will last as long as language. Yet the idea that this cellular theory was the only explanation of immunity was quite short-lived. Indeed, men like Ehrlich, Behring and Roux were having other opposed ideas while Metchnikoff was still formulating it. The alternative explanation, which looked more promising, became known as the humoral theory of immunity; its central argument was that microbes are killed by chemicals in the blood. Behring, when he produced his antitoxin against diphtheria, had discovered *antibodies*. These are substances in the blood, manufactured by the body to deal with *antigens*, which can be invading viruses, or practically any other things that intrude on the body's functions.

It was obvious even then that immunology was a highly complex subject. Neither the cellular nor the humoral theories were entirely satisfactory. However, the appearance on the scene of an Englishman of Irish descent did much to promote a widespread interest in, and a basic understanding of the subject. He was Sir Almoth Wright, or 'Sir Almost Right' as his many not imperceptive critics called him. True to the one nation's reputation for compromise when the occasion suits, and the other's for occasionally promoting irrational argument as a method of arriving at the truth, Wright proposed a link between the two emerging theories of immunity.

Wright was at once a maddening and endearing character. He was never other than an intensely human scientist. He could be dogmatic and rude, or gentle and considerate, depending on whether it was to his advantage to demonstrate any of these qualities. He was almost theatrically puritanical in his attitude to food and drink, and blatantly prejudiced against women. A long letter he wrote to *The Times* opposing votes for women is said to have caused the whole of the Irish vote in Parliament to swing against universal franchise.

But Wright had plenty of compensatory virtues. He had, like so many of those of Irish descent, a useful command of the English language and once thought of taking up literature instead of medicine. The result was that he successfully and unashamedly popularised the work he was doing in newspaper articles, not giving a fig for the fact that the contemporary doctor-doyens of his trade thought it criminal that the hoi polloi should be made to understand in simple terms what medicine was all about.

The man who probably perpetuated Wright's popular fame was George Bernard Shaw, who modelled his character, Sir Colenso Ridgeon, in *The Doctor's Dilemma* on the doctor scientist, and gave the phrase 'stimulate the phagocytes' to Edwardian hypochondriacs. Shaw had been visiting Wright's laboratory when one of Wright's assistants happened to say that yet another patient had applied for treatment that afternoon. Shaw then turned to Wright and asked him what would happen if more people applied for help than could properly be looked after. In his dogmatic fashion Wright answered, 'We should have to consider which life was the best worth saving.' Shaw put his index finger to his nose and said, 'Ha! There I smell drama – there I smell

149

drama.' The doctor's dilemma had crystallised. When he went to see the play, Wright walked out. He thought it a travesty of his true beliefs and work.

In 1892 Wright had become professor of pathology at the Army Medical School at Netley on Southampton Water and had begun to work on typhoid fever with the intention of finding a vaccine. To the British, with an Empire intact, typhoid fever was imperially important. Not only were there thousands of deaths from it in England each year, it also killed off large portions of the army in distant, hot and insanitary places such as India. Until now success with vaccination had been based on the use of living bacteria, though Pasteur's greatest victories had been with attenuated, or weakened, varieties. In order to eliminate altogether the very real risk of a patient catching typhoid from the bacteria, no matter what weakened state they were in, Wright decided to break with the Pasteur tradition and kill the micro-organisms responsible by heating them, then use them as a vaccine. His results were encouraging. In Wright's own somewhat subjective estimation they were downright incontrovertible.

Wright instituted more quantitative methods of making and using his vaccines than had ever been thought necessary before. He accurately estimated the numbers of bacteria present in his vaccine solution and he also related the dose he used to the response of the antibodies in the patient's blood. This work, as with most things he dabbled in, attracted controversy. He had made a great advance by showing that in some cases safe vaccines made from dead bacteria could give immunity. But though Wright had no doubt about the efficacy of the vaccine, the Army – staff, officers and men – were divided. As with all newly introduced vaccines there were set-backs – sickness suspected to have been brought on by the vaccine itself – and some deaths. Some military authorities were not convinced by Wright's statistics on the preventive value of his typhoid fever inoculation; nor for that matter was the distinguished statistician Karl Pearson, who, with a well-argued case, fought a considerable battle with Wright in the British Medical Journal over Wright's interpretation of his own results.

Although the vaccine had been ready for the outbreak of the South African war, and Wright had been allowed to use his very considerable powers of persuasion on the medical authorities to solicit volunteers, army tradition and the universal maxim 'never volunteer' had been against him. He had more success in India when he visited regiments there and was allowed to stand face to face with his potential patients, and address the troops himself. But still by 1914 typhoid vaccination

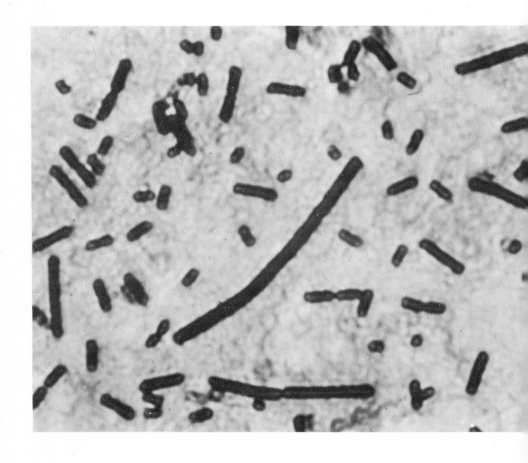

was not compulsory. Wright had to tackle Lord Kitchener before an order was placed making all soldiers travelling abroad have inoculation against typhoid fever. Even so, in the Great War 20,000 men contracted the disease and 1100 died from it. The figures compare well with the Boer War when a much smaller army saw 50,000 of their numbers succumb to it and 9000 die from it: more than were killed by Boer bullets.

Wright's contribution to resolving the cellular and humoral theories of immunity was to propose that the blood contains substances which in some way affect the phagocytes' mode of action and allow them to deal with the attacking bacteria more effectively. Wright (one of whose hobbies was to invent new words) called these substances *opsonins* – from the Greek *opsono*, I cater for; I prepare food for. The opsonin, he supposed, prepared the bacteria-victuals more effectively for the phagocytes.

The therapeutic methods which Wright based on his theories, inoculating his patients with vaccines to stimulate the body's own immune mechanisms, had a popular vogue which lasted many years. But they were a blind alley down which research workers had to retreat. He did, however, stimulate other things in his laboratory at St Mary's Hospital where, from 1903, he headed a research department. One product of that laboratory was Alexander Fleming's work on penicillin; Leonard Colebrook's work on puerperal fever also had its origins there. Wright died in 1947, sometime before the rise of women's liberation, but not before he had published *The Unexpurgated Case Against Women Suffrage*.

It was shown eventually that the opsonins act on micro-organisms and not on the phagocytes. These reactions are, however, only a small part of the complex chemistry involved in immunity. It took half a century for chemists, who had long suspected the existence of antibodies, to hunt them down as the chemist Pasteur had hunted down the microbe. A microbe that operates as a vaccine is one that has had its structure changed in such a way that its ability to damage cells is impaired, but which can still stimulate the formation of antibodies to deal with the invading antigens.

It took time to learn the skills needed to produce a range of anti-viral vaccines. In 1937 Max Theiler succeeded in making an attenuated yellow-fever virus which could be used to give complete protection against the disease. In the early 1950s Jonas Salk produced a very effective vaccine against polio in which killed viruses were used. But even here, in relatively modern times, there occurred a terrible tragedy which again stirred up public reaction against the principle of vaccination.

When wide-spread polio-vaccination was being introduced in 1955 in the United States some improperly prepared samples of the vaccine actually produced the disease in many children – by now almost a predictable repetition of the experiences of Jenner, Pasteur, Wright and many others. But again, in spite of the initially tragic shortfalls, the technique survived. In 1957 Albert Sabin produced a vaccine against poliomyelitis prepared from a living virus which could not itself cause polio, but which could cause antibodies to appear to deal with the effects of the live attenuated polio virus.

These methods cause what Paul Ehrlich called 'active immunity' in man. The vaccines bring about the production of antibodies and the effect of the vaccine is long-lasting. 'Passive immunity' is the protection given by the short-term measure of feeding antibodies produced from animals into the human body. A well-known example of a passive immunological agent is gamma globulin, which gives protection for a short time against measles.

How the human body produces the right antibody to deal with the invading antigen at the right place at the right time is a remarkable and remarkably reassuring process. It appears that the body has its readily available range of antibodies which it can call on when it meets a new antigen.

The superb ability of the human organism to behave in this way does cause man to face very serious problems when, for example, he tries to manipulate it into some adventure in which it has no mind to indulge. Like it or not, everybody is allergic to everybody else's body. If a piece of skin, or a heart, or a kidney, is taken from one body and transplanted to another body, the entering tissue is treated as foreign. Immediately the mechanism by which antibodies are produced is called into play to fight off and reject the new tissue. Unless special techniques are used, this immune response will succeed; and even when every effort is made to suppress it the protective mechanism of the body may still have an easy and cold victory. The costly unfulfilled hopes and the sad outcome of heart transplants in the late 1960s and early 1970s testify to the problems still to be overcome.

Towards the end of the nineteenth century there were physicists who were ready to forecast that their subject was circumscribed; the major problems in it had been solved and before the end of the century it was suggested, after the tying up of a few loose ends, the science could be wrapped up in a neatly arranged parcel as a symbol of man's intellectual competence in one area of natural philosophy. These physicists were soon to be robbed of their arrogant illusion. The Einsteins, the Rutherfords, the Bohrs and the Schrödingers showed both what a mere smattering of the language of physics had been translated, and what a vast tome was left to be read and understood.

Life scientists did not make the same mistake as their physicist counterparts. The reason might well be that the very ignorance man has of so much of his own body is daily too troublesome to let him fall into ridiculous assumptions about his own omniscience. Immunology is as noble a challenge as was bacteriology a hundred years ago; then, the subject-matter of immunology was inconceivable since a knowledge of bacteriology and the life sciences had not revealed it. In a century's time immunology will have spawned many new and important subjects which today are outside our conception. By that time we shall probably have solved the problem of how a body recognises what is itself, and why it rejects what is not itself. It is possible that a way will have been found of making our immune systems so controllable that some constituents of our own bodies can overthrow cancer cells. And it is possible that the elegant mystery of why a pregnant mother does not reject her own foetus will have been equally elegantly explained.

Detail of a painting showing Pasteur and Lister (on steps) jointly honoured by the Sorbonne in 1892

CHAPTER TWELVE

In the 1840s pregnant women in the Vienna General Hospital were to be seen on their knees begging not to be taken into one of the obstetric clinics because they rightly feared that in that clinic they would stand a good chance of dying of childbed fever. Not much more than a century later in that city, and in other European cities, the number of deaths from puerperal fever was negligible. Between these two extremes lies the story of the development of the germ theory – a theory which no serious evidence of the last century has shown might be disproven. It is a magnificent piece of scientific creativity with sweeping social consequences which turned a significant part of medicine from a mystical mess into a rational craft.

The process of science which led to the creation of the whole was made up of many small, but identifiable creative acts. The words 'theory' and 'proof' imply a methodology. If this piece of 'creativity' was of such profound importance, would it not be useful to analyse the steps involved in its creation in the hope of repeating them with some other socially significant end in mind? Any such analysis must of course look closely at the methods of biology, physics and chemistry with which the characters involved occupied themselves at their laboratory benches. But there are also other more subjective elements that contribute to creativity; there is an artistry in science, and it shines out from the work of many of the people connected with the germ theory. To probe the nature of their work demands that one probes the characters and motives of the men themselves.

There are some obvious common characteristics among the chief architects of the theory – Pasteur, Koch, Lister, Metchnikoff, Ehrlich and the rest. That they were all capable of the rational thought that is the chief exercise of scientific activity goes without saying; without this faculty they could have made no mark whatever on science or its applications. Yet many lesser men with the same easy grasp of logic have failed to make any dent where these men left signally clear impressions. Another necessary characteristic, and one which those who have never been inside a laboratory nor seen the process of scientific research in action might either forget or discount, is that all of them were obsessively involved with their subject; they were prepared to devote the greater part of their waking lives to routine, repetitive, often boring and sometimes difficult laboratory procedures. What they did was hard work: crystallising, precipitating, culturing, observing. For hour after hour and for day after day they would be willing to repeat the same tasks, knowing that the majority of their experiments would show no positive result.

There can be no doubt of the obsessive natures of so many people attracted to scientific research. Pasteur wore his on his sleeve. He was once seen walking up and down the rows of cages of animals kept for experimental purposes, tearing down the labels stuck on each one by his assistants, and replacing them with cards written in his own hand. He desperately needed it to be seen that every facet of his personality – even his handwriting – was clearly visible on the work in progress. On one research trip in Africa, Robert Koch shook awake his pupil, Friedrich Kleine, at five o'clock in the morning to tell him that a horse had just died from a suspected sickness they were investigating, and needed dissecting immediately. Kleine, who was worn out by Koch's demands, protested that he was too tired and that he had been dreaming about horses all night. Koch was quite taken aback that his assistant's obsessions were not identical with his own. 'How can you hope to make any progress,' he asked the astounded Kleine, 'if you don't dream about horses?' And yet other men have had deeper obsessions, and worked many more hours over their culture plates and microscopes than did either Koch or Pasteur, but have produced nothing whatever that remains in anybody's memory.

Classifying men according to type is a hazardous business at the best of times, but a glance at the psychology of some of the characters in this tale is provocative. The proportion of hypomanics – those who fly off into bouts of energetic physical and mental activity and then sink into periods of profound depression – is striking: Jenner, distracting himself with his cuckoos to try to drag himself from his fits of depression; Semmelweis, storming past the close ranks of beds in his clinic into pointless arguments with his superiors; Metchnikoff, twice driven to attempting suicide, and lifting himself out of his trough with technicolor dreams of scientific fantasy. There are others; but it would need a Freud rather than a statistician to make a convincing case from so small a sample. The psychopathological mechanism involved in examples such as this may well be, as some psychiatrists have argued, that the manic depressive seeks to win love, approval and acclaim with short aggressive bursts of activity: hence the creativity. The theory does nothing to explain the mechanism of creativity, nor can it easily be falsified. But the characters of scientists of the germ theory do help sustain it.

A more universal characteristic of the same group of men was their willingness to be involved in (or their inability to prevent themselves from becoming involved in) an argument. It is almost possible to find a correlation between the argumentative nature of the man and the num-bef and quality of his ideas. Pasteur is the prize exhibit. His rows, which when convenient he engineered in public, stimulated his creative abilities like a drug. The more witnesses there were to the act of indulgence, the better he seemed to like it. He was at his most humanly destructive, and at his most scientifically creative, when he was able to parade his giant intellect over that of, say, the unhappy Colin at a crowded Académie de Médecine; the theatrical gesture of dumping an anthrax-infected hen on the desk in front of his startled fellow Academicians gave to creativity that extra sweetness that makes it all so much more worthwhile.

Koch, the other Master, indulged in the same practice, though with neither the same relish nor skill as Pasteur. Behring loved to argue. Ehrlich hated it. Yet they both did it frequently, and sometimes with one another. However, there is some difficulty in making argument into a precursor of creativity. It is difficult to separate cause from effect. It could just as easily be supposed that science attracts argumentative types to its laboratories, as to assume that the processes of science convert those who sample them into an argumentative disposition.

A study of the character types of the microbe hunters tells little enough about creativity, though it is reassuring on one point. That is that the men who have sat at laboratory benches and made some of the most impressive discoveries in history do not begin to fit into a single character pattern that can be called 'the scientific type'. In the bodies and minds of Joseph Lister, Ignaz Semmelweis, Elie Metchnikoff, Emile Roux and the rest were displayed all the strengths and the weaknesses, the anger and the sentimentality, the egoism and the humility, the astuteness and the idiocies of any other cross-section of educated men.

Their motives, however, are as interesting as their types, and more definably a spur to creativity. The first motive that needs to be dealt with, because it is the one most frequently and most inadequately argued as a reason for the involvement of so many astute minds in the formulation of the germ theory, is that these men were moved to help suffering humanity. That their work did make once indescribably dreadful diseases into conditions that need not now be feared, is without question. Also there is little doubt that when they saw the results of their work in operation – when Pasteur saw Joseph Meister recover both from the deep bites of the rabid dog, and from the pin-cushion punctures of Grancher's vaccinating needle; when Ehrlich watched a syphilitic ulcer disappear after an injection of salvarsan – then the rich feedback from the results was deeply rewarding. Many of the men

159

themselves, and certainly their biographers, attribute this desire for man's well-being, 'the health and happiness of the world', as one put it, to have been the sole purpose of their work. But the evidence that exists suggests that in most cases this was an entirely retrospective motive: a result of the work and not the cause of it. Pasteur arrived at his work on human diseases at the end of a long series of interrelated intellectual exercises which had had as their applications, among other things, stereochemistry, the improvement of vinegar yields, the prevention of silk-worm disease and the establishment of a more secure wine industry for France, and better beers for customers of England's Whitbreads. And when Koch first set out on the trail of the anthrax bacillus, his first move was to the intellectual haven at the back of his consulting room and away from the suffering patients who day after day turned up at the front.

Undoubtedly, some chose the subject of their researches because they felt a need to relieve suffering, which was therefore a prime motive for their work. Joseph Lister fits into this category and so too perhaps does Elie Metchnikoff. Both these were deeply affected by pain and unhappiness. However, Metchnikoff's most significant work, which reached a peak of creativity when, in his beautifully simple experiment, he took a rose thorn and stuck it into a starfish larva, stems from his interest in zoology as much as it does from a concern for the human condition. His least successful work was that in which he tried to think up ways to 'correct the disharmonies of human nature'. It is almost a measure of his failure to find any practical solution to this problem over which he fretted so much that in the end he turned to an artist, Tolstoy, for solutions.

As an idealised motive, the relief of suffering is a fine thing. However, some of the challenges that moved these men to tackle the intellectual hurdles were much less praiseworthy, much more powerful and, to those who are attracted by the psychological condition of man, much more interesting. The germ theory – a tricky scientific obstacle race if ever there was one – drew out every competitive characteristic in those who took part. Pasteur's ambitions were made even more interesting by the fact that he wanted not simply to win ownership; he wanted to be *seen* to be in possession. The bigger the crowd that applauded, the more secure he felt in that ownership. The applause sealed the title-deed and ensured him the right to say, 'This is my own, my very own.'

The barricades Koch built round himself, and the tradition of secrecy inherited by his assistants, produced unseemly squabbles in which any pretence at a scientific brotherhood was discarded. But science is a very human activity. To pretend that all is rational and reasonable within its perimeter is to misunderstand its very nature.

But when it came to applying the consequences of the germ theory to medicine, still less admirable motives became apparent. There was money to be made. Emil Behring liked money, and he liked what it could buy him: his villas on Capri; his castle in Marburg, his private laboratory and (to the satisfaction of those who did not hold him in particularly high esteem) his mausoleum. There is no reason to assume that the profit was a less powerful spur to him than it was to the Henry Fords and the Thomas Edisons. And in the longer term the effects of competition and of the profit motive on the drug industry which sprang from the germ theory have been remarkable – if dispiriting to the idealist. The sulphonamides, streptomycin, chloramphenicol – indeed virtually all the revolutionary drugs – were developed and marketed in Western capitalist countries. Communist state-run drug industries have no list which can compare.

Not all profits made out of the germ theory were on a vast scale, nor intended for use on conspicuous luxury. Perhaps Robert Koch *was* propelled into the tuberculin fiasco in an attempt to finance an estranged wife, Emmy, and a mistress, Hedwig. But even if that were proved to be the case, there is something not altogether damnable and even a little touching about the elderly scientist looking for a way of legitimately financing some personal contentment for his later years.

Louis Pasteur made nothing from his bacteriological successes. He firmly adhered to the Franco-British tradition that scientific purity was incompatible with personal gain. Nevertheless, he had a substitute for the profit motive which was no more admirable: a lifelong and passionate nationalism. His love of France and his hatred of Prussia verged on the paranoic. Pasteur, in his later years, saw his own work as being humane and therefore good, in contrast to the work of those involved in warfare and violence on behalf of his country's enemies. Ideologically, he was convinced about his nation's cause and her science's part in that cause: 'We may assert that French Science will have tried, by obeying the law of Humanity, to extend the frontiers of Life.'

But there is one impulse which singles out man from all other creatures: the sheer pleasure to be derived from creation for its own sake. And there can be no doubt that, in their different ways, the microbe hunters did derive indescribable pleasure from the intellectual exercise of the search for the truth. Sir Almoth Wright used to blurt out his anticipation of that pleasure each day as he walked through his laboratory. 'Well friend,' he would say as he passed an assistant, 'what have you won from Mother Science today?' And he would listen for the reply, hoping to cap it with an achievement of his own. And what prevented Metchnikoff from throwing himself to his death on that depressed evening on the Rhône bridge was the realisation that there was some inexplicable satisfaction to be had from explaining the mystery of the evolutionary history of a passing mayfly.

But the anticipation of pleasure, like every other motive to create, gives only obscure clues as to how creativity can be systematised and repeated. More substantial clues should be available in methods. A scientist can use any method he likes in his search for the truth. There are no limitations, and no rules. Traditionally there is an order of events which a scientist should follow in his search. He should closely observe Nature and what it has to reveal, then induce the laws of Nature from what he has seen. Paul Ehrlich claimed that he worked in this fashion. As he put it, he would refuse 'to give directions to Nature but simply try to analyse striking experimental facts which were difficult to understand and by so analysing to find the laws which governed the action'. Ehrlich, however, was notoriously self-contradictory about his methods of working and was capable of producing an alternative mechanism if it seemed to explain better the origins of one of his discoveries. One of his most astounding pieces of creative science was to make the extraordinary assumption that the textbook formula of atoxyl was incorrect. On the basis of this guess – it was nothing more – and on the basis of the guess that derivatives of atoxyl might be capable of acting like rounds of ammunition on a target of microbes, salvarsan and the science of chemotherapy were given birth. To be sure, it took 605 chemical compounds and thousands of experiments before salvarsan came to light. But the creative leap had been made when Ehrlich plucked these two basic ideas out of the air and *then* worked out the experiments needed to verify the ideas.

This method, of forming a hypothesis, and then devising experiments to verify the hypothesis is by far the most common way for scientists to carry out their work. What characterises the maturation process of the germ theory is the number of quite remarkable pieces of intuition which contributed to the whole great edifice. And it is the intuitive act which is at the nub of the creative process. The greater the man, the more extreme and unorthodox the intuitive act appears at first sight: Pasteur – preventing Chamberland from throwing away the chicken cholera culture that had stood undisturbed during the summer vacation, because of a guess that something had happened to the microbes during that time; Ehrlich – with enough knowledge of chemical reactions to give him confidence in his guessed structure for atoxyl and to be willing to see two of his best chemists walk out on him because of it.

In any group of men which includes intellects as singular as those of Koch, and scientific talents as unusual as those of Ehrlich, it is difficult to argue a case for there being any one man having exceptional genius. But there is something in the affairs of Louis Pasteur which, however indefinably, shines a shade more brightly, with a little more passion and an extra face of beauty. Moreover, Pasteur did not simply perform one act of scientific creativity – or even two. He performed them throughout his life and to order. He of all the men in this brilliant group might be expected to give us some glimpse of the true nature of creativity: some hint of how the act might be repeated.

Now it is true that there *is* a repeatable pattern to the way he approached each new problem. First he made sure that he had clearly defined the problem. Without prejudice, he saw what had already been done, and sifted through as much published work as was available. Then after a vitally important gestation period of what might in some cases be minutes, and in others years, he made the imaginative leap – he formed his theory. Finally proceeding with great skill and practical knowledge he tested the theory. The pattern can be followed from the moment when he first looked down his microscope at a tartrate salt to his last vaccination against rabies.

Pasteur was well aware of the crucial nature of the creative hypothesis at the centre of each piece of work. He admitted that he always worked from preconceptions. Some of these he wrote down as notes long before he had any experimental evidence to substantiate them. His method was shared by some of the most revolutionary and successful minds to have worked in science. Gauss, the mathematician and physicist, was once asked how soon he thought it likely that he might come to some conclusion about a difficult mathematical problem. He replied that he had already come to his conclusion; what worried him was how to reach it. Einstein was explicit on the same point: 'A theory can be proved by experiment, but no path leads from experiment to the birth of a theory.'

Pasteur had his share of luck: so do all successful scientists. The random event was of importance in the life of Louis Pasteur (as it was in those of Paul Ehrlich and Alexander Fleming) because accidents that happened in his laboratory gave some meaning to his prior conceptions. Pasteur, in addition to his just share of luck, had flair – a quality which seemed to be lacking in so many of his contemporaries. He had a sense of theatre which gave his scientific creations the character of personal performances. No discovery of Pasteur ever got within a whisker of passing unnoticed. In the days before 'charisma' was a fashionable word, Pasteur dramatised his work – built up suspense then brilliantly directed his final scene. The final act of publication whereby a piece of research becomes a formative part of the current of scientific knowledge is essential. Without it, it ceases to exist. So firmly did Pasteur throw his work into the public arena that he might have dispensed with publication. Less histrionic figures must use the now established means of communication, publication in a scientific journal, to make their mark.

But for all this analysis, are we any closer to understanding the mechanism which produced the essence of Pasteur's scientific intuition? Do the tantalising, quick visions which appear when, for an instant, the veil seems to blow away from the source of his inspiration, tell us anything useful? Or are they chimeras: fanciful things we imagine because we feel the need to understand?

Pasteur's methods show consistent, repeated patterns; his character was plainly hewn and his motives were clear and fiercely stated. The study of this man makes an observer feel far closer to a mechanistic explanation of creativity than with any other of the microbe hunters – great men all of them, but lesser than Pasteur. Yet, though he brings us so close to the border that we feel sure he can tell all, what we are looking for, though infinitely more clearly definable and admirable as a result of his life's work, remains elusive.

But if, in the end, any analysis is unsatisfactory and if we cannot formalise any conclusions, other than to observe that the highly successful methods adopted by Pasteur were repeatedly creative, we can at least note the conditions under which creativity flourished and usefully ask whether these are the conditions being given to the Pasteurs of our own age.

Pasteur's scientific imagination was unconventional. The more unusual or even outrageous were his intuitive hypotheses, the more profoundly influential and disturbing they were to the path of scientific knowledge. Unconventional thoughts are not conceived by conventional minds. But the fields in which Pasteur had his most striking successes – bacteriology and immunology – were not those in which he had had a formal conventional training. No such training existed. His education had been as a chemist. There are many other examples of highly creative scientists who, having started life, say, as physicists, and having turned to biology as mature men, have then produced their best and most revolutionary work.

Whereas it cannot be denied that a great number of crucial scientific discoveries have been made by those whose education has followed a conventional pattern to turn them into part of the scientific élite, it must be asked whether or not some minds are constrained or blinkered by this pattern. As a result of the competitive process which produces the élite, how many minds are being excluded from making radical contributions to scientific creativity? It is a well-known fact in engineering companies that people without scientific training, taken from the production line and asked to work in research laboratories alongside research engineers, can, after a few months, suddenly flower in a remarkable manner. Their naïvety prevents them from accepting conventional barriers to thought and they often develop once inconceivable talents for engineering creativity. They share with Pasteur the advantage of an able mind, and of having brought it to the subject of their research unstultified by a collection of preconceived ideas.

Therefore, rather than so readily accept that it is desirable that students of science be encouraged to indulge in specialisation for substantial periods of their lives, why should not a deliberate act of scientific educational policy be one in which young people are positively encouraged to change direction in mid-stream? Why should not conditions be made such that it is easy for them to do this without fear of being eliminated from the competitive educational ladder? There is no evidence that this would be less creatively productive than the original direction, and there is considerable evidence that occasionally revolutionary ideas will emerge which, besides being of value to a scientific culture, give satisfaction and fulfilment to the individual.

It has never been possible to focus properly on the character of creativity: always there is inexactitude and ambiguity. A mystery remains whose understanding, philosophers of science maintain, is irrelevant to the understanding of the process of science. This might be so; but man will try to penetrate the mystery for the same reason he tries to create. And he will feel closer to the edge of the mystery if he can reproduce the conditions in which the creative act can flourish.

CHAPTER THIRTEEN

During their lifetimes the microbe hunters were elevated in the eyes of the general public to the status of heroes. Arguments over whether the good of mankind was or was not a motive for their work seem irrelevant in the face of the ghastly suffering which their work eliminated before the end of the nineteenth century. In the Middle Ages gunpowder and plague had together destroyed the social structure of Europe. After the evolution of the germ theory the fear that plague might play the same devastating role yet again was removed. In the 1850s potato blight effectively reduced the population of Ireland from eight to four millions. Once the reason why a nation could be part-obliterated so tragically was understood, the first sure step towards preventing the recurrence of such a disaster could be taken.

At Pasteur's birth the average life-span in European countries was about 40, and in some communities as low as 25. At his death it was nearer 70. He died before the award of the first Nobel Prize in 1901. But by 1908 Behring, Koch, Ehrlich and Metchnikoff had all been given the prize for medicine. It gave them fame and money to match the esteem which few grudged them. One correspondent, writing in *The Spectator* in 1910, having made a pilgrimage to Pasteur's grave noted that above it were carved four great white angels: Faith, Hope, Charity, and Science. Pasteur and his colleagues had thus raised science to the same level as the three great Pauline virtues.

Today these same men are still looked on as heroes because their work seems so clearly to be life-enhancing and destined, if left to run its natural course, to produce a better world. They are heroes even at a time when public attitudes to science and scientists have altered dramatically. The last quarter century has seen much science – above all, physics – become suspect. The ability of a small élite group of men, not dissimilar in some ways to the microbe hunters, to produce an atomic weapon and see it through to use in warfare caused many outside science to pause and wonder. When the possession of nuclear weapons and their threat began to dominate international politics, the worst fears of many were realised. Science, which had once seemed to be only a power for good, had produced a weapon which was now threatening to be as socially disruptive as gunpowder had been in the Middle Ages, and one infinitely more frightening because of the enormity of its power to destroy.

If Pasteur, Koch and Ehrlich are the heroes of an earlier scientific age, who then are the anti-heroes of our time? Enrico Fermi, who brought about the first chain reaction in uranium? Robert Oppenheimer, who organised the élite scientific group that produced the first atomic bomb? Edward Teller, the father of the hydrogen bomb? It is difficult, if not impossible, to see any fundamental difference between the motives and the methods of this group of men and those of that earlier group who discovered weapons with which to fight disease. The atomic physicists were as clever, as modest, as self-seeking, as mean, as argumentative and just as concerned for humanity as the microbe hunters. The physicists worked to produce a weapon of war, but there is no real evidence that the nationalistic arguments which convinced them that their efforts were right and just were any different from those that so affected Koch and Pasteur half a century earlier; and the intellectual challenge was just as great, and grappling with it just as enjoyable. The scale of the physicists' operation was greater, but their methods were neither more nor less easily definable than those of the biologists; nor was the process of creativity any less beautiful.

The physicists' work was widely seen as being culpable because it was applied to the taking of life; the first two atomic bombs did so on a vast and horrifying scale. But it was Pasteur, and not some atomic physicist, who in 1870 said of the Germans, 'I want to see the war prolonged into the depths of winter, so that all those vandals confronting us shall perish of cold and hunger and disease.' And it was Georg Gaffky who, in 1916, said: 'We owe it to Koch that our armies are free of epidemics.' In the Franco-Prussian War French army doctors performed 13,000 operations of which there were no fewer than 10,000 fatalities. The germ theory made it possible for the next war between the two countries to be more prolonged, and with more deaths in rather than out of battle.

In the short term the germ theory saved lives and was therefore admirable. But any analysis of such an influential field of science must look to its long-term effects. There were some early worries in the first years of the century about the moral implications of work on microbes. For example, was it desirable for society, some asked, that chemotherapeutic research had turned up with a cure for syphilis and gonhorroea? On balance the feeling that it was a good thing overcame the notions that the wicked should be left to suffer Nature's own terrible punishment. Even well into the 1950s it was argued that the mere understanding of the nature of the contagion was enough to reduce the spread of venereal diseases. Penicillin, however, proved to be such an effective therapeutic agent that it removed the great fear that had once waited on victims, and by the 1960s and 1970s epidemics of gonhorroea in Western European countries were a more serious cause for concern than they had ever been.

The use of chemotherapeutic agents over not very long periods of time produced some disturbing problems. Initially man had quick triumphs over bacteria and disease, but soon Darwinian selection began to operate in favour of certain micro-organisms.

From the 1940s to the present day there has been a continuing change in the susceptibility of bacteria to the many antibacterial agents available. The simple act of exposing a bacterium to a drug favours the survival of a resistant strain. Those who were fighting microbes had to develop new weapons; what had once been forecast as a march to victory turned into an endless war of attrition, with research workers jumping from one predicament to another. Nevertheless, the extensive and cruel scourges that once ate at man's existence have disappeared. The scenes of suffering that were, not so many years ago, a frequent sight in all European cities have gone. Smallpox and the plague are no longer the common killers. Instead, the diseases of affluence, cancer and heart ailments, have taken their place. Human beings are kept alive longer, and eventually succumb to an increasing number of metabolic and degenerative disorders, rather than to infectious diseases. The human lifespan has been extended, but not necessarily the useful or enjoyable human lifespan. Conquering the common diseases has produced vast and, at present, insoluble problems of distress and suffering in old age.

The one enormous problem to which the germ theory of disease has contributed, by virtue of its success, is that of population increase. This increase has spawned and aggravated other problems: the drain on the earth's resources, pollution, spoliation of the environment and the physical and psychological sufferings of overcrowding. For some years now many have regarded these effects as the harvest of uncontrolled or irresponsibly applied science. The new concern over environmental problems is the most recent result of increasing public disaffection and distrust of science and scientists. That there is cause for concern is indisputable, but in looking for solutions to these mounting problems it would be unwise to overlook the causative part played by the science whose aims were, on the face of it, entirely admirable. It is one of the greatest paradoxes of this century that what Western civilisation now looks on as major evils – over-population, pollution and the rise in the incidence of diseases of affluence – can have their origins traced to the work of founders of the germ theory as easily as to any other scientific source.

For the non-scientist to condemn the work of physicists or chemists as irresponsible, and to put the microbe hunters on a pedestal, would be to misunderstand the process of science and the nature of human involvement in science.

During the second half of the nineteenth century the germ theory affected the course of science in much the same way that impressionism affected painting, or Beethoven's last symphony had affected music fifty years earlier. It can be argued that art and music have influenced nations: that a Goya satire changed the course of politics, or that a Mozart opera spurred a revolution. But in truth, these works of art were surface symptoms of hidden illnesses in the societies in which they were created. Scientific creativity is not a surface manifestation of some submerged social unrest or stimulus. When it occurs it has inevitable and continuing consequences for which some share of the responsibility must rest on the shoulders of the scientist: the creator. Whoever believes that the responsibilities carried by the microbe hunters were less, or their errors smaller, than latter-day scientists, is blind to the phenomenal effects of the germ theory. As with all science it had facets of beauty as indefinable as all aspects of beauty. As with all science it had sharp edges to inspire fear because, as with all science, it was capable of being applied, and the applications resulted in evil as well as in good. Whoever applies science and controls it controls power. The germ theory of disease turned out to be a wholly unsuspected force whose immensity was only realised during the twentieth century when other scientific sources of power were also being made manifest. By this time mankind was learning to be more fearful of the side-effects of that power when it ran out of control. One thing is certain: these troublesome effects – pollution, energy crises, radioactive fall-out, diseases of affluence, and all other spectres which to some typify the twentieth century – can only be combated by scientific means: by the process of hypothesis, deduction and rational argument that make up scientific method. No other solutions worth considering, from whatever source, have been or are likely to be proposed. It is all the more important therefore that, in order to use science to this important and laudable end, and to control its products for our own good, we should use every means to understand the genesis of science. Merely because what we are seeking is elusive is not sufficient reason for abandoning the attempt to analyse the brushstrokes, or the counterpoint or the inspiration of science which give rise to the creative scientific act.

Acknowledgement is due to the following for permission to reproduce illustrations on the pages listed:
Ad Makeup Co. Ltd., 56; F. R. Elwell, 147 (top left); Mary Evans Picture Library, 116, 121; The Mansell Collection, 12, 21 (top left), 22 (top), 40, 51 (top right), 81, 90 (top), 95 (top right), 124, 156; Radio Times Hulton Picture Library, 21 (bottom), 51 (top left), 63 (bottom), 66 (top), 102, 143 (both); Ronan Picture Library, 8, 21 (top right), 51 (bottom), 63 (top and centre), 86 (bottom), 89 (bottom), 95 (bottom), 166; St Mary's Hospital Medical School, London, title page, 134 (both), 139 (both), 140; Dr Valentine, Medical Research Council, 147 (top right); C. James Webb, 22 (bottom), 66 (bottom), 86 (top), 89 (both), 147 (bottom), 152; The Wellcome Institute of the History of Medicine, 26, 95 (top left), 151.